OIL AND THE MIDDLE EAST WAR:
EUROPE IN THE ENERGY CRISIS

HARVARD STUDIES IN INTERNATIONAL AFFAIRS
Number 35

Oil and the Middle East War: Europe in the Energy Crisis

By

Robert J. Lieber

Published by the
Center for International Affairs
Harvard University

Harvard University
Center for International Affairs
Executive Committee

Created in 1958, the Center for International Affairs fosters advanced study of basic world problems by scholars from various dis-
ciplines and senior officials from many countries. The research of the
Center focuses on economic, social, and political development; the
management of force in the modern world; the problems and relations of
advanced industrial societies; transnational processes and international
order; and technology and international affairs.

The Harvard Studies in International Affairs, which are listed at the
back of this book, may be ordered from the Publications Office, Center
for International Affairs, 6 Divinity Avenue, Cambridge, Mass. 02138,
at the prices indicated. Recent books written under the auspices of the
Center, listed on the last pages, may be obtained from bookstores or
ordered directly from the publishers.

About the Author

Robert J. Lieber is Associate Professor and Acting Chairman of the Department of Political Science at the University of California, Davis. He received his Ph.D. from Harvard and has held an International Affairs Fellowship from the Council on Foreign Relations and a Guggenheim Fellowship. He has also been a Research Associate at the Harvard Center for International Affairs as well as a Visiting Fellow at St. Antony's College, Oxford, and at the University of Essex. Professor Lieber is the author of *British Politics and European Unity* and *Theory and World Politics*, and co-author of *Contemporary Politics: Europe.*

Contents

OIL AND THE MIDDLE EAST WAR:
EUROPE IN THE ENERGY CRISIS

INTRODUCTION

The European energy crisis — exacerbated by the October 1973 Middle East war — comprised a series of dramatic and at first sight unique developments: a sudden awareness of political and economic vulnerability as a consequence of lopsided dependence upon Arab oil, an oil embargo of the Netherlands and cutbacks for other members of the European Community, shifts in policy toward the Middle East, simultaneous pressures for European integration as well as for fragmentation, a scramble for bilateral agreements with oil-producing states, massive balance of payments costs, increased inflation and decreased economic growth, and an oscillating and sometimes disruptive pattern of relationships between Europe and America. The crisis also involved both traditional national actors (the countries of Europe, the Middle East, and North America) and several transnational ones (the European Community, international oil companies, OPEC, the OECD, the IMF, and the newly created International Energy Agency). Yet for all this, the case of Europe in the energy crisis is by no means *sui generis*. Its economic and political elements illustrate major tendencies with a wider application,[1] and the manner in which Western European governments and the European Community have confronted crucial oil and energy problems provides data for a series of fundamental questions about the effects of transnational and traditional international processes on regional cooperation and conflict.

A central question of this study is to determine why the European Community failed to cohere when faced with the energy crisis. Prior to the Yom Kippur War it was widely assumed that the members of the EC might well respond to the stimulus of a pressing external challenge by presenting a common front. The Community, newly expanded to nine members with the addition of the United Kingdom, Denmark, and Ireland, had come to be regarded as a major actor in international affairs, particularly in their important economic dimensions. In reality, the Nine[1a] responded to the crisis in disarray, dividing their actions among three major alternatives: 1) efforts at a common *European Community* response, as a means of dealing as a bloc with the Arabs and the U.S.; 2) narrowly national policies of *sauve qui peut*, aimed largely at *bilateral* deals with individual oil producers; and 3) a *multilateral* course of action involving cooperation among the principal developed industrial states, especially the U.S., and creation of an International Energy Agency (IEA) to coordinate their actions in dealing with oil producers.

This study begins by sketching the background to the crisis, noting

1

the shift of Europe from a coal to an oil-based energy economy and the series of abortive European efforts to create a coherent energy policy prior to the October war. It then deals with the major elements of the European response to the energy crisis, particularly the way in which the Europeans found themselves pulled in three directions simultaneously. Next, it analyzes the reasons why Europe emerged from the initial stages of the crisis, in the words of the EC Commissioner for energy policy, "weakened and humiliated."[2] Among the principal causes examined are the dramatic politicization of the energy issue, the increased role played by political differences among the Nine, and the way in which the energy crisis re-established the Europeans' vulnerability in security terms — an area that had previously seemed of diminishing importance to a world in which issues of international economics had moved to the fore.

Finally, in an afterword, the analysis concludes by using the elements of the above case to develop five propositions. These concern the politicization of international economic relations and the consequences of this for policy responses based on national, European, and multilateral levels.

1 BACKGROUND TO THE EUROPEAN ENERGY CRISIS

European integration has advanced more rapidly in the technical sectors than in the political ones. To be sure, this distinction is best made on the basis of how something is treated rather than its inherent nature,[3] but the European Community clearly has achieved its most substantial results in the fields of agriculture, tariff policy, internal free trade, and common commercial policies, while it has made little progress in establishing common political institutions or coordinating foreign policy and defense. Yet the seemingly technical field of energy policy had not experienced substantial integration prior to the onset of the October 1973 war. Indeed, energy remained an institutional orphan.

Ironically, given the absence of a common European energy policy, energy (in the form of coal and nuclear power) figures prominently in two of the basic European Communities: The European Coal and Steel Community (ECSC), established in 1952, and the European Atomic Energy Community (Euratom), established in 1958. Other energy resources, including oil and gas, lie within the purview of the European Economic Community (EEC, or "Common Market"). Nowhere, however, do the treaties establishing these institutions call for an overall common energy policy nor do they pose a timetable for its future creation; indeed, the division of responsibility among the ECSC, Euratom, and the EEC tended to work against any such measures.[4]

By failing to provide for a Common External Tariff on coal and other energy imports, the ECSC Paris Treaty helped to insure the decline of the European coal industry and the movement of Europe from a coal-based to an oil-based energy economy. The Treaty dealt with problems of cartels and initially banned coal subsidies, but otherwise left states free to set high tariffs to protect their coal production in competition with cheaper imported coal, or to import foreign coal — and then oil — as economically as possible in order to provide lower-cost energy supplies for local industries. In general, countries which relied primarily on imported energy sources obtained energy cost advantages over their ECSC partners. The effect was to distort their comparative advantage because of the relatively high proportion of energy in the final cost of finished industrial products — e.g., 26% in steel, 16% in chemicals.[5] In addition, national tax policies and pricing strategies of the major oil companies had the effect of favoring the use of crude oil at the expense of coal. Taxes and retail prices of gasoline, which did not compete with coal, were kept high, while crude oil was dumped on European markets at prices from one-half to one-seventh of those of gasoline.[5a] Under these con-

3

ditions, European manufacturers made increasing use of cheap foreign coal and then cheap imported oil. In 1950, approximately 75% of Europe's primary energy requirements had been met by coal and only 10% by oil; a decade later the ratio had slipped to 61% versus 33%; and during the 1960s the relative positions of coal and oil in Europe became reversed, so that by 1970, coal provided only 33% and oil 60%.[6]

Europe's vulnerability to interruption of its energy supply was foreshadowed by the 1956 Suez War. With the temporary closure of the Suez Canal, the Organization for European Economic Cooperation (OEEC) established a Petroleum Emergency Group within which representatives of European governments, the OEEC, and oil companies met to arrange a reserve pool for oil. This permitted some sharing of oil with those members facing supply problems.[7]

As the competitive position of coal worsened,[8] there were attempts to grapple with the energy problem. In 1962, European coal producers urged subsidization of coal production to prevent future vulnerability to oil supply and price problems. In April 1964, the ECSC adopted a "Protocol of Agreement" on energy policy, which called for broad objectives of cheap and secure supplies of energy, freedom of choice for consumers, and fair competition among energy sources. These vague guidelines contained a major contradiction: energy supplies could not be both cheap *and* secure. Either security of supply would be given priority through enhanced support of European coal production — at relatively high prices, or cheap foreign supplies (especially oil) would become increasingly dominant, thus leaving Europe vulnerable to interruption of supply. The 1964 Protocol did leave room for state subsidies of coal, but — in contrast to agriculture — made no provision for a shared community financing scheme, nor did it provide for specific production figures. In essence, the ECSC thus opted for a policy of cheap fuel. Those states with sub-regional problems involving declining coal industries provided their own national protection or subsidies, others imported coal and oil at cheaper world prices. In the mid-1960s oil replaced coal as Europe's primary energy source.

Following the Six-Day War of June 1967, a number of Arab states undertook an oil boycott of some pro-Israeli European states and the U.S. Ample alternative supplies were available from other oil producing areas (North Africa, Nigeria, Venezuela) and a degree of consumer coordination took place via the Paris-based OECD (which now included the U.S. and Canada as well as most of Western Europe). The oil embargo failed completely, making far less of an impact in Europe than the 1956 post-Suez shortages, and it was soon terminated by the Arab producer states.

4

In December 1968, the EC Commission produced a series of recommendations in its "First Guidelines for a Common Energy Policy." This document reaffirmed the 1964 objectives of cheap and secure supply, freedom of consumer choice, and fair competition between diverse energy sources. The 1968 Guidelines went somewhat further in specific proposals, recommending support for European coal by means of certain import restrictions and a subsidy program (through national funds). In addition, the Commission sought diversification of oil import sources, improvements in data collection and forecasting, the establishment of a common market for energy (by removing non-tariff and tax barriers within the EC), and the creation of stockpiles equal to 65 days' oil consumption. While the Commission expressed concern over nationalistic tax, supply, and marketing policies and the resultant energy cost disparities distorting industrial competition, its own recommendations remained nebulous. The EC Council of Ministers approved the "Guidelines" in principle in 1969, but with the important exception of stockpiling, which was implemented, national government policies remained divergent, particularly for coal, for which no actual import limits were established. Ironically, the Commission had continued to reject the idea of any contradiction between security of supply and low price:

> The Community's interests demand first and foremost security of supply at prices which are relatively stable and as low as possible. There is no foundation for the assertion that this requirement is a contradiction in terms.[9]

By 1970, EC energy policy thus amounted to very little beyond a 65-day oil stockpile requirement and a series of vague guidelines. Member states pursued divergent policies toward coal, some protecting their local mines and others relying mainly on imports of cheap foreign coal from the U.S. and Eastern Europe. There existed no common market in energy, and in general Europe followed a policy of low-cost supplies based on huge imports of cheap oil from the Middle East and North Africa. (See Table 1.) Nor had Euratom achieved any success in joint atomic energy policies. By this time it had proved a virtual failure by reason of national rivalries and reluctance to relinquish control over sensitive nuclear research. This had led to fragmentation and duplication of effort, with Euratom increasingly mired in bureaucracy and scientific stultification.[10]

By this time, the balance of competitive advantage between consumers and producers had begun to shift. The wide margin of excess world oil production capacity had diminished, and OPEC became an in-

TABLE 1

Primary Energy Consumption by
Form of Energy, OECD Europe

	Percent	
	1960	1970
Solid Fuels	61.4	29.4
Oil	32.5	59.6
Natural Gas	1.8	6.7
Hydroelectricity	4.2	3.3
Nuclear Energy	0.1	1.0

Source: OECD, *Oil, The Present Situation and Future Prospects* (Paris, 1973), p. 265.

creasingly effective vehicle for coordinating the producers' policies. In 1971 and 1972, OPEC succeeded in achieving substantial price rises and by the end of 1972, crude oil from the Persian Gulf was selling at 60-70% above its mid-1970 price.[11]

Ironically, in view of the Community's later disarray, the Commission did see in advance the major dangers to which Europe was exposed by its overwhelming dependence on Middle East oil imports and the implied vulnerability to both supply and price changes. In October 1972, the Commission submitted a lengthy memorandum on energy policy, with forecasts and recommendations for the period to 1985. This time the Commission conveyed a sense of urgency, particularly on the problem of oil supplies, which it saw in the future "threatened by more or less widespread interruptions."[12] Finally, the Commission grappled with the supply versus price trade-off by advising that "long term security" be given priority over temporary price advantages. It recommended talks with the U.S. and Japan to improve OECD procedure on information and stockpiling as well as the establishment of 90-day stockpiles and the adoption by member states of laws making it possible for them to cope jointly with future supply difficulties. At the same time, the Commission proposed consultations with oil-exporting countries on the basis of a complementarity of interest extending beyond the energy field. It recommended agreements to promote the oil exporters' economic and social development in exchange for guarantees by these producers on the export of oil to the EC.[13]

Foreshadowed in these recommendations were two divergent

courses of action: cooperation with other consumer countries, particularly the U.S. and Japan via the OECD, and special arrangements between the European Community and the oil exporters. The inability to make a choice between these courses of action was to prove damaging to the Community when the crisis began a year later.

The October 1972 Paris summit meeting of the now-expanded nine-member EC gave support to a common energy policy but, beyond agreeing to 90-day stockpiles and encouraging efforts to guarantee supplies at satisfactory prices, the Paris summit left most of the details to be worked out by the Commission for later submission to the Council of Ministers. Thus in April 1973 the Commission produced its "Guidelines and Priorities for a Community Energy Policy." It contained some obvious recommendations (development of local energy sources, increased use of coal, steps to achieve a decrease of 10% from anticipated oil imports in 1985); most significantly, however, the Commission advocated not only development of relations between Europe and the producers, but also cooperation with the U.S. and Japan in order to avoid "needless and expensive overbidding for crude oil by the importing countries."[14] The "Guidelines" also called for steps to cushion the impact of any crisis by sharing oil imports if any member state were affected[15] and by establishing in advance a joint energy consulting group with the U.S. and Japan, via the OECD.

Remarkably, the Commission had set out the criteria by which the Community's choices and shortcomings could be gauged in the crisis half a year later: avoidance of overbidding, sharing of oil among member states, consumer consultation with the U.S. and Japan, cooperation with producers. A month later, in May 1973, the Council of Ministers held its first meeting on energy policy in three years. In essence, the Council failed to reach any substantive agreement on consumer collaboration or emergency energy sharing. The French government refused to authorize talks by the Community with the U.S. and OECD, on the grounds that the Nine must first reach agreement on their own policy and that talks with outsiders could only be conducted by the governments, not by the Commission. Sharp disagreements between the French and their EC partners were papered over by the Council instructing the Commission to make concrete energy proposals before the year's end. These differences reflected sharp underlying political differences. Externally, France aimed to improve her position against the historic domination which Britain and the U.S. had established in Middle East oil after World War I. She thereby sought special arrangements with Libya, Algeria, and other producers. In addition, a refusal to cooperate closely with the U.S. reflected French preferences for enhanced political and economic

7

TABLE 2

Primary Energy Needs of
The European Community, 1973

	Million ton oil equivalent	Percent
Solid fuels	227	22.6
Oil	617	61.4
Natural gas	117	11.6
Hydroelectric power and other	30	3.0
Nuclear energy	14	1.4
Totals	1005	100

Source: European Communities, Commission. *Towards a New Energy Policy Strategy for the Community.*
(Communication presented to the Council by the Commission on June 5, 1974). See *Bulletin of the European Communities*, Supplement 4/74, p. 12.

autonomy *vis-à-vis* America, both for France and if possible for the Nine.

Mid-1973 was a period of extreme European-American antagonism; France resisted Secretary of State Kissinger's design for a new Atlantic Charter in this "Year of Europe" and sought to defend her interests via European means when possible. Bitter European reaction had already occurred over U.S. monetary policies and temporary restrictions on American soybean exports and there was thus no consensus on the proposed cooperation among energy-consuming nations which the Americans had proposed at the OECD in 1972.

Thus on the eve of the Yom Kippur war, Europe remained extremely vulnerable in the energy field. Oil, virtually all of it imported (two-thirds of it from Arab countries), provided 61% of Europe's primary energy supply. (See Table 2.) Despite the obvious need for a more coherent regional response to the energy problem the member governments resisted agreement on specific Commission proposals because the Nine had differing priorities. There were political disagreements about the extent to which Europe should cooperate with other consumers, particularly the U.S. Some support existed for the French idea of a modus vivendi with Middle Eastern and Mediterranean oil producers based on both Europe's geographic location and high need for imported oil and upon a desire for a more autonomous European world role. Others favored broad consumer cooperation based on the recognition of America's huge advantage in energy resources, nuclear power, and even military might, in order to deal with the producer cartel. Economic perspectives also differed: Britain, the Netherlands, and the German Federal Republic followed relatively laissez faire policies in dealing with the multinational oil companies, while France and Italy were highly *dirigiste* in their organization of the internal market. Britain and the Netherlands were the home base of two of the major oil companies (BP and Shell) while France and Italy sought to build up the position of state-owned companies.[16] In addition, member states faced divergent sub-regional problems: Britain, for example, had a declining coal industry which required subsidies in order to maintain production and employment; others, such as Italy, had industries which depended on imports of the cheapest coal and oil available. Even geology played a part in obstructing easy agreement. Britain and Germany obtained roughly one-third of their energy from coal and the Netherlands more than two-fifths of its requirements from local natural gas. They were thus less dependent on oil imports than the French and Italians, for whom oil represented about three-fourths of energy needs. (See Table 3.) And there was even the prospect of Britain's North Sea oil wealth

TABLE 3

Proportional European Community Dependence on Primary Fuels
(1973)

	Oil	Percent Natural gas	Coal	Other
Italy	78.6	10.0	8.1	3.2
France	72.5	8.1	16.1	3.2
Belgium Luxemburg	62.1	13.8	23.7	0.4
Netherlands	54.2	42.3	3.4	0.1
West Germany	58.6	10.1	30.1	1.3
Britain	52.1	13.2	33.6	1.2

Source: British Petroleum Statistical Review of the World Oil Industry, 1973, in Louis Turner, "Politics of the Energy Crisis," *International Affairs* (London), Vol. 59, No. 3, p. 408.

which promised to alter radically her situation by the end of the 1970s.

In view of these deep differences, the failure of the Community to reach a common energy policy prior to October 1973 is not surprising. But a series of statements by European leaders throughout the first half of 1973 proclaiming the need for and possibility of "European Union" (Brandt[17]), a "European foreign policy" (Heath[18]), a "European identity" (the Council of Ministers[19]), and even European defense cooperation (Jobert[20]), seemed to hold out the possibility that the obstacles to agreement might well be overcome in the future as a political consensus developed on the need for a more unified and coherent European role.

2 EUROPE IN THE ENERGY CRISIS

Necessities are the great federators, at least in the opinion of Jean Monnet. And, by mid-1973, substantial reasons existed to support the assumption that the European Community countries might well respond to pressing external challenges by increasing their degree of integration. This orientation was typified by a then current quip that if Europe did ever achieve unity, statues should be erected to the two men most responsible for its success: Joseph Stalin and John Connally. Indeed, there was already talk of an energy crisis and some foreign policy elites were convinced that it would intensify pressures for common European policies.

Before the onset of the Yom Kippur War, Europe was widely regarded as increasingly a major actor in international affairs. Richard Nixon had enshrined the Europeans as one of five principal participants in an emergent world balance of power system ("I think it will be a safer world and a better world if we have a strong, healthy United States, Europe, Soviet Union, China, Japan, each balancing the other, not playing one against the other, an even balance.").[21] To be sure, this picture contained more fiction than reality: Europe was not at all united enough to act as a coherent entity in such a balance and it remained militarily dependent upon the United States. There was much else of dubious analysis in this balance of power conception, enough to deprive it of any descriptive or predictive validity.[22] Yet there remained a sense in which Europe did constitute an economic power of the first magnitude, expressed for example in Zbigniew Brzezinski's notion of two triangles of power, an economic one in which Europe joined Japan and the U.S., as well as a military triangle consisting of the U.S., the USSR, and China.[23]

However, as already noted, the reaction of the European Community and its nine members to the energy crisis was one of substantial disarray. The pre-existing policy disagreements — reflecting divergent outlooks on energy, the Middle East, Atlantic relations, and European supranationalism — as well as Europe's resource vulnerability and lack of sufficiently coherent political institutions, made a common and sustained European response difficult to achieve. The Community responded to the crisis in fragmented and sometimes contradictory ways at successive stages, dividing their actions among three major alternatives: first, efforts at a unified Community response — including the option of dealing as a bloc with the Arabs and with the U.S.; second, strictly national policies, particularly in efforts to obtain bilateral deals

11

with individual oil producers; third, a multilateral response involving cooperation among the principal developed industrial states, especially the U.S.

a. Response as a Community

At the outbreak of war on October 6, 1973, the European Community still lacked a common energy policy and its members remained divided in their attitudes toward the Middle East and the U.S. Aware of their vulnerability, the Nine did however seek to avoid involvement in the conflict, and this quickly brought them into sharp disagreement with the U.S. Indeed, in the first days of the war, Britain's Prime Minister Heath even refused to support an American proposal for a cease-fire resolution at the UN. But the sharpest antagonisms developed over the American resupply of Israel. Most of the Nine refused overt cooperation in this effort; thus, on October 10, Heath's private request for a "cover story" on the use by U.S. reconnaissance planes of a British base in England or Cyprus provoked Secretary of State Kissinger's rage and led to a temporary ban on the exchange of American intelligence information with Great Britain.[24] The Federal Republic at first quietly allowed U.S. arms shipments to Israel via Germany, but after a highly publicized report appeared in the German press on October 24th, showing an Israeli ship loading arms at Bremerhaven, the German government asserted that its neutrality required the cessation of such operations. Italy also refused basing rights, and France manifested its pro-Arab policy by continuing to ship tanks to Libya and Saudi Arabia.[25] The subsequent American "Defcon Three" military alert on October 25th added to European-American differences. Heath and President Pompidou of France refused to cooperate with the U.S. effort to block a Soviet move to include both Soviet and American troops in an international emergency force to supervise the cease-fire, and French Foreign Minister Jobert later bitterly criticized direct U.S.-Soviet dealings which accompanied the termination of hostilities.[26]

A few days after the outbreak of war, the initial joint response of the Nine to the conflict was a call by the Council of Ministers for a cease-fire and negotiations based on the November 1967 UN Resolution No. 242. This produced little response, partly because of the divergent positions of the Dutch and the French. But on October 17th, the crisis suddenly hit the Europeans directly as the Organization of Arab Petroleum Exporting Countries (OAPEC) announced its embargo. In its essentials, the OAPEC measures, as further elaborated on November 5th, involved successive monthly cuts in oil production until such time as Arab objectives were achieved in the conflict with Israel. Countries were divided into three categories: those, such as the U.S. and the Netherlands, which were to receive no Arab oil, "friendly" countries, including Britain and France, which were to receive normal supplies based on previous 1973 delivery levels, and other countries, which would face phased reductions of 5% per month. At a stroke, three different kinds of treatment were meted out to members of the EC.

By this time the French government of President Pompidou had begun to call for a coordinated European position toward the Middle East, so that the Nine might prove "their capacity to contribute to the settlement of world problems."[27] The immediate European response was a "Statement on the Situation in the Middle East," which essentially supported the Arab interpretation of UN Resolution 242 and which stressed both Israeli withdrawals and formal recognition of Palestinian rights. Although this document has received divergent interpretations, as both European "self-abasement"[28] and as scarcely differing from previous and widely accepted UN resolutions[29], the context in which it appeared was substantially one of appeasement of the Arabs. Indeed, it angered both Secretary Kissinger and Egyptian President Sadat by calling for greater Israeli concessions than Sadat was seeking at the time.[30] This shift toward the Arab position produced an immediate, though minor, payoff: on November 19th, OAPEC oil ministers decided not to impose their 5% cutback on European Community oil scheduled for December — except of course for the Dutch.

Thus, in the initial stage of the crisis, the Nine had managed to achieve not only a semblance of common policy toward the Middle East, but to derive some tangible benefit. Nonetheless, as the question of the Dutch embargo became more overt and the need to choose between siding with the United States (to grapple with problems of supply, price, conservation, and monetary recycling) or OPEC (to seek a modus vivendi with the oil producers) became more acute, European policies were rapidly reduced to disarray.

The embargo of oil for the Netherlands by OAPEC presented an unusually clear — and uncomfortable — choice for the EC partners of the Dutch. The Arabs regarded the Netherlands as pro-Israeli because of positions taken by the Dutch government. These included a statement by the Dutch Foreign Minister that Syria and Egypt had "broken unilaterally the coexistence maintained since 1970,"[31] an October 7th request that the belligerents return to the 1967 cease-fire lines (which the Arabs incorrectly interpreted as Dutch endorsement of these as final borders), and an earlier offer by the Dutch government of a substitute transit center for the Austrian Schonau Camp which had been used by Soviet Jews but closed after a terrorist incident. By contrast, the French

13

and to a lesser extent the British, Belgians, and Italians, adopted more pro-Arab positions. Thus, while OAPEC embargoed all oil to the Dutch (the same treatment accorded the U.S.), the French, British, and later the Belgians were rewarded by being placed in OAPEC's "friendly" category.

The Dutch response, as expressed by Foreign Minister Max Van Der Stoel, was that EC principles must be adhered to in guaranteeing equitable fuel supplies for all members.[32] The EEC's Rome Treaty provided for free circulation of commodities within the Common Market, thus any attempt by the Eight to honor or cooperate in the embargo would violate the Treaty. On October 30, 1973, the Dutch requested EC agreement to pool oil supplies and share them with the Netherlands if this should become necessary. France, fearing to jeopardize her Arab ties, almost immediately rejected the request, Prime Minister Messmer replying with casuistic logic that the precondition for such sharing was a common European energy policy — which the Nine lacked.[33] The British government was less explicit but similarly negative in its position.

On November 6th, the EC Council of Ministers, while issuing a mildly pro-Arab statement — which the Netherlands endorsed — refused to take the actions requested by the Dutch; instead it produced a bland and meaningless declaration that it was conscious of the interdependence of members' economies, and asked the EC institutions to "follow attentively" the situation resulting from the scarcity of oil.[34] Despite the fact that the Arab oil boycott violated a 1970 UN General Assembly Resolution (No. 2625) — for which the Arabs had voted — forbidding the use of economic or other measures to coerce other states;[35] despite the earlier expressed willingness of public opinion in the EC to support the principle of member countries coming to the aid of another member in the event of an economic or energy crisis (ranging from 59% approval in Britain, to 70% in France, to 88% in Italy);[36] and despite the various commitments contained in the Community Treaty, the Nine were unable or unwilling to present a publicly united front on behalf of the Dutch. It is likely that a position of European solidarity would have enjoyed firm public support and a considerable probability of success, yet with the notable exception of Germany, most other member governments either felt themselves too vulnerable in their dependence on imported Arab oil or gave priority to maintaining or seeking a modus vivendi with the Arabs. In this sense, the energy crisis was considered a matter of national survival: oil supplies were crucial for industry, agriculture, heating, electricity, and transportation, and narrow conceptions of national self-interest received priority over European solidarity. Economic interdependence and the existence of regional integration could not — at least temporarily — override notions that oil had become a matter of national security.

The inability to cooperate was not confined to the nine-member EC; previous European oil supply problems in 1956 and 1967 had been dealt with through the more broadly based OEEC and its successor, the OECD. While no formal system for sharing existed within the Community, the OECD did possess such a framework in its Oil Committee (a special group of oil company officials). This body met on November 21-22, but — lacking the necessary unanimity — decided against authorizing its International Industry Advisory Board to activate its allocation system.[37] By most accounts, it was French and British refusal which blocked OECD action.

Although the basis of bargaining among the Europeans had thus shifted to one of traditional national interest calculation, the Dutch were not without resources of their own. Beginning on November 15th, the Netherlands government issued a series of warnings to its EC partners that if oil imports were to drop sharply and the Community failed to respect its own rules, then the Netherlands might be forced to curb exports of natural gas to France, Belgium, and Germany. Since 40% of France's natural gas consumption — and much of the supply for cooking and heating in the Parisian region — came from Dutch fields, this was no trivial matter.[38] From this point on, a quiet resolution of the Dutch situation appears to have taken place. On November 20th, the Dutch expressed satisfaction over an unspecified "common position" reached among the Nine to ease the embargo. A week later, a high official in the French Foreign Ministry observed that the selective action against the Netherlands could not force France to set herself apart from her Common Market partners."[39] Meanwhile, it had become evident both that the United States government would encourage the international oil companies to aid the Dutch and that the companies would be likely to undertake the task of allotting the reduced supply of oil among the consumer countries. The Dutch government, while maintaining that, in the words of Prime Minister Joop den Uyl, it "would never place the continuation of the State of Israel in jeopardy,"[40] began to ease its position toward that of its EC partners by pledging that no transport of weapons or volunteers to Israel would be based in Holland, and subsequently by characterizing the Israeli occupation of Arab territories as "illegal."[41] While the boycott continued for six more months, its practical significance declined from the early December point at which the Dutch Prime Minister had described the decrease in oil to the Netherlands as 30%.[42] In early January, the Dutch and the Arabs actually agreed to

15

allow Arab oil to pass through the huge Dutch port of Rotterdam en route to other European countries, particularly to Belgium, which had been placed on the Arab "friendly" list on December 26th. The Dutch themselves began gasoline rationing on January 12th but ended it on February 4th. While most OAPEC members terminated their embargo of the United States in mid-March, they continued to boycott Holland. The Dutch responded by threatening to block EC initiatives toward the Arab countries, and even French Foreign Minister Michel Jobert delivered an elliptical criticism of the embargo.[43] Finally, in mid-July, the Arabs lifted their embargo. By that time, the port of Rotterdam was already operating at 80% of capacity; oil for domestic use had regained its normal level in April and in June Dutch crude oil reserves actually exceeded their pre-embargo figure of October 1973.[44]

The highly political nature of the oil embargo with its attendant differences in national priorities and policies had overridden the transnational forces of economic interdependence and regional integration by precluding overt agreement on EC or OECD oil sharing. However, one major group of transnational actors, the international oil companies did alleviate the crisis and lessen the potential for conflict among the consumer countries by allocating the oil themselves. With at least the acquiescence of the European governments (who also feared to antagonize the Arabs), and (according to testimony by the then head of the Federal Energy Agency, John Sawhill, to the U.S. Senate Subcommittee on multinational corporations) at the urging of the U.S. government, the companies acted to allocate the available oil on an essentially *pro-rata* basis. In the words of an executive of Royal Dutch/Shell:

> The allocation of oil as a percentage of demand to all markets appeared to be the most equitable and practicable course of action in the circumstances. Indeed it was the only defensible course if governments were not collectively to agree on any alternative preferred system.[45]

Specifically, Shell claimed to have impartially cut all its markets by 17%, which was comparable to Arab reductions as a percentage of world trade in crude.[46]

In essence, this allocation procedure represented — in game theory terms — a kind of "prominent solution." In the absence of overt agreement, it provided the least disagreeable outcome for the major participants. The logic and acceptability of this outcome, as well as the strength of the oil companies in implementing it is reflected in the fact that although most European governments sought to improve their own

16

supply situation, they experienced similar supply reductions.[47] Thus Prime Minister Heath had a bitter clash with executives of the partly state-owned British Petroleum Company over allocation priorities and the French government ordered the CFP (also partly state-owned) to divert oil from Japan to France. Robert Stobaugh notes that while the French situation proved an exception to the pattern that oil companies did not discriminate in favor of their home markets, nonetheless even France lost about the same share of its oil as did the other countries.[48] One compelling explanation for this outcome is that neither Britain nor France persisted too firmly when the companies resisted their pressures and replied that the governments would have to announce publicly which countries would have to suffer shortfalls if the companies were forced to discriminate.[49] Indeed, as early as December 11th, the oil companies told the French government that despite its privileged position France would receive 10 to 15% less oil at the end of the month.[50] Yet it was not merely the British and French governments who tried to exact national advantage: Italy sought full crude oil deliveries from the companies and restricted exports of refined products, and Belgium and the Netherlands temporarily licensed or restricted exports. The companies, however, threatened that crude oil would be diverted elsewhere if petroleum product exports to the U.S. and other destinations were blocked.[51] They thus exercised considerable muscle of their own, for example, by later boycotting Belgium in order to force the government to allow higher domestic prices.

The end result was that regardless of their standing with the Arabs, whether in the "friendly" category (Britain, France, Belgium), "neutral" (Germany), or embargoed (Holland), the members of the European Community received roughly comparable percentage cuts in their pre-crisis petroleum supplies. In this sense, the oil companies, acting in a highly interdependent oil market of ports, supplies, refineries, pipelines, and trade in refined products, succeeded in diverting oil from non-OAPEC suppliers and in producing a tolerable outcome. At the time of the greatest shortfall, U.S. petroleum supplies available daily were 7.4% below the September level, and consumers in the rest of the world (particularly Europe and Japan) averaged a loss of 6.7%.[52] These effects were manageable; the greater impact of the crisis became evident not as a problem of supply[53] but one of price.

The most striking result of the embargoes and production cutbacks had been to drive up the price of crude oil from $2-2.50 per barrel until it ultimately stabilized at approximately $10. From October 1973 to January 1974, this meant a 400% rise. The price increase received its greatest impetus not only from the Arab producers and from Iran and

Venezuela, but also from wild competitive bidding for scarce supplies. Indeed, in February 1974, the French company ELF-ERAP signed a long-term agreement to buy Libyan crude oil at $16 per barrel (a price which Valérie Giscard d'Estaing, then Finance Minister, later forbade the company to pay).[54] By late January 1974, the EC Commission predicted the direct and indirect 1974 impact of the oil price rise as likely to cause inflation of an additional 3%, economic growth of only 2-3% (1½% below average), increased unemployment of 0.7%, and a net balance of payments cost of $17.5 billion.[55] And for Britain, Italy, and France, subsequent estimates predicted overall 1974 payments deficits of $4-8 billion each.

With the oil boycott in effect, the EC held its first summit meeting since the start of the energy crisis. Gathering at Copenhagen on December 14-15, 1973, the nine heads of state or government confronted a series of difficult choices: support the Dutch or placate the Arabs, formulate common EC policies or pursue divergent national responses, stress efforts to reduce oil prices or concentrate on security of supply, explore cooperation with the U.S. and other consumer nations or follow a predominantly European approach. The governments of Germany and Denmark sought firm common EC policies in support of the Dutch, even to the extent of sharing oil supplies. The Heath and Pompidou governments, however, emphasized the need for a special relationship with the Arab world, involving both a diplomatic effort to restore oil supplies and intensified trading relationships, including a triangular arrangement in which excess Arab oil producer revenues would be used to subsidize Egypt which would buy European exports, thus providing Europe with the recycled funds with which to buy still more high-priced oil.

In the course of the Copenhagen summit, five Arab foreign ministers made an unexpected and dramatic entrance to the conference hall where they spoke to urge intensified pressures against Israel. This unprecedented intervention, arranged unilaterally by the British and French, reflected the extent to which the Arabs' use of oil had left the Europeans weak and divided. Ironically, the opportunity to take part in European consultations before EC policy was finally determined had long — and unsuccessfully — been sought by the United States.

In their summit communiqué the Nine formally reiterated their declaration of November 6th, expressing support for the Arab interpretation of UN Resolution 242, but cautioning the Arabs against the possibility of a negative reaction of European public opinion if the oil supply reductions continued. The fundamental disagreement among the EC members produced little more than a reference to the need for Europe to speak with one voice and a statement on energy which ignored both the embargo of the Netherlands and any suggestion of a joint stand against the oil producers. In the absence of immediate agreement, the Commission was assigned the task of drafting proposals for the orderly functioning of the energy market, increased efficiency in energy use, the development of alternate energy sources, measures for research and development, and proposals for cooperation among *both* producers and consumers — and these proposals were to be presented to the Council of Ministers before the end of January 1974.[56] For the Nine, the outcome of the summit was thus meager: it may have contributed to the willingness of the Arabs to provide increased supplies of oil,[57] but at a high price, while producing no more than vague declarations on dealing with producers and consumers and on the future development of common policies.

The Copenhagen summit made increasingly explicit the inability of the European Community to produce a common policy to deal with the energy crisis in either its broader international dimensions or in its internal aspects. Indeed, far from providing the necessary stimulus to new steps toward policy cooperation, or institutional development, or the enhancement of common attitudes, the transnational effects of the energy crisis intensified the divisions within the Community. On the one hand, the crisis presented a set of intractable international problems involving the Middle East, Atlantic relationships, security, Europe's world role, supplies of oil, and distortions of trade and payments flows. These issues necessarily intensified the relevance of basic and unresolved political differences and priorities among the Nine, not only in regard to policy but over EC institutions as well. At the same time, differences among them were exacerbated by domestic problems within the principal member states. In each case, the international aspects of the energy crisis reached into national arenas, where they posed increasingly intractable problems of employment, inflation, energy use and allocation, fiscal and monetary policy, and sharp disagreements among political parties, elites, interest groups, and the broader publics over preferred policies both internal and external.

In France, increased pressures from a reinvigorated Left weakened Pompidou's political base and left the French president more susceptible to Gaullist strictures against greater cooperation with the U.S. and more willing to conclude bilateral Arab deals. As far as the French government was concerned, if the Community could be induced to follow these lines of policy as a group, so much the better, but the payoffs from closer EC cooperation in the absence of such agreement were insufficient to outweigh the domestic costs.[58]

19

Germany showed growing resentment over the increased costs of the Community, particularly in financing the Common Agricultural Policy (CAP) for the benefit of France. By the end of 1973, Germany had already paid out $3.5 billion net to other EC members; without real political or institutional progress toward greater European unity, further costs, for example, in financing an EC regional policy to benefit Britain, Italy, and Ireland, would be difficult to justify domestically. In addition, Germany's national security interests made her the most pro-NATO of the EC members and thus unwilling to follow the Arab-oriented policies of Britain and France. These currents intensified with the resignation of Willy Brandt and the accession to the Chancellorship of Helmut Schmidt in April 1974.

Finally, the British government faced mounting internal problems. The long awaited membership in the Community had brought neither immediate and tangible economic benefits nor a silencing of domestic antagonisms over the costs and benefits of EC entry. At Copenhagen, Prime Minister Heath thus resisted suggestions that Britain's North Sea oil would be of particular benefit for her partners. Balance of payments and other problems also led his government to seek to improve some of the terms of Britain's membership, particularly by securing a substantial regional program to aid depressed British manufacturing and coal-mining regions. German unwillingness to finance these costs on a sizable (multi-billion dollar) scale led the Heath government, on December 29th, to block any further steps toward an EC energy policy, thus stalemating the Community in its plans to hammer out the broad range of common policies (regional, social, industrial, energy, agricultural, etc.) called for by the October 1972 Paris summit.

These factors dominated the attempts to arrive at common energy policies within the Community and to deal with OPEC and the growing U.S. efforts to assemble a broad-based front of the developed consumer countries. In December, shortly before the Copenhagen summit, Secretary Kissinger had suggested consumer cooperation to the Europeans; on January 3, 1974, he warned against the development of bilateral dealings with the oil producers, and on January 9, the U.S. formally invited the leading European states: Britain, Germany, France, and Italy, plus Norway and the Netherlands (as well as Japan and Canada) to a Washington energy conference on February 11. The invitation coincided with the French announcement of an oil deal with Saudi Arabia, involving a three-year agreement to exchange arms, industrial products, and technology for 27 million tons of oil.

In response to both these developments the EC Commission issued its own statement urging a joint European reply to the U.S. offer and cautioning against the danger of bilateral dealings between individual governments and oil producer countries. This position was amplified by the Commission in its January 31 report to the Council of Ministers. Noting that the problem of oil price rather than supply had now become central, the Commission expressed concern over a "real danger" that EC members might pursue mutually inconsistent and damaging economic policies in dealing with balance of payments deficits, protectionism, contradictory fiscal policies, competitive devaluations, and capital movements, and insisted that these problems could not be overcome at the national level.[59] The Commission also issued an unusually explicit statement that Europe was in "a state of crisis," which involved serious "setbacks and failures" for the nine members and the EC institutions, and it observed prophetically that the forthcoming Washington Energy Conference would pose a crucial "test" of the EC members' ability to cooperate.[60]

In anticipation of the conference, the Commission had sought an agreed EC response to the American initiative, cautioning against individual national approaches which would weaken the "political solidarity the Community intends to demonstrate as evidence of its existence and significance."[61] In mid-January, the Community countries had agreed that all nine of them would go, in effect forcing the U.S. to extend invitations to Belgium, Luxembourg, Denmark, and Ireland, with the Community itself to be represented by Walter Scheel, then German Foreign Minister and President of the Council of Ministers, and Commission President François-Xavier Ortoli. On February 5, a week before the conference convened, the EC Council of Ministers reached agreement on a joint position (Britain having lifted its veto on energy policy in response to progress on the regional issue). The nine foreign ministers adopted a policy which lay in the direction advocated by the French: the Washington Conference was not to establish a confrontation against the oil producers, no new permanent organization was to be created, there must be a commitment to prompt discussion (i.e. by April) with less developed countries and the producers, and work emanating from the conference was to be carried out through existing bodies such as the OECD and IMF. Within these limits the Nine were willing to discuss problems of supply, price, oil sharing, and public control. The Council also reached agreement on close governmental monitoring of oil movements and prices — a policy advocated by France, moving away from previous reliance on the oil companies for this information. Responding to these signs of enhanced European policy coordination, Commission President Ortoli exclaimed that the U.S. had pushed the Nine into more European unity on oil in two hours than had been achieved in the previous ten months.[62]

21

As the conference opened in Washington on February 11, 1974, Scheel and Ortoli articulated their common positions, giving particular attention to the need for broad-based measures of cooperation to avoid competitive devaluation, protective tariffs, and overbidding for oil. In stating the European view Scheel also stressed the need to transcend a merely regional context in dealing with the crisis.[63] However, the common position of the EC members began to disintegrate as bitter antagonisms appeared between the French and the Germans. These differences, which had been temporarily papered over, predated the conference. Indeed, the very idea of the conference had interfered with French plans for a special European-Arab relationship, and this had led Michel Jobert to characterize the U.S. proposal as a "provocation."[64] By contrast, Helmut Schmidt, then Foreign Minister and leader of the German delegation to Washington, had criticized the "go-it-alone" policies of France and Britain.

In essence, Schmidt and the Germans agreed with Kissinger's stress on the need for an agreed code of conduct limiting bilateral deals, and on the desirability of broad measures of consumer cooperation based on a recognition of interdependence. Kissinger expressed the choice directly: "Will we consume ourselves in nationalistic rivalry which the realities of interdependence make suicidal? Or will we acknowledge our interdependence and shape cooperative solutions?"[65] France basically rejected Kissinger's stress on a multilateral approach and interpreted the prior Council of Ministers' position as supporting her refusal to cooperate. The Eight, led by Germany, viewed the Council position as flexible enough to allow support for specific proposals advanced by the U.S. These differences culminated in a dramatic personal confrontation between Schmidt and Jobert. The German Finance Minister argued that the energy crisis was too big to be managed on an exclusively European basis and pointedly invoked political and military linkages, noting that by themselves "The Europeans . . . can't even maintain a balance on their own continent,"[66] and adding that only the U.S. had the power to get and keep a settlement in the Middle East.

The final communiqué of the Washington Energy Conference embodied most of what the U.S. had sought. Not only was there agreement on a commitment to conservation, demand restraint, the development of alternate energy sources, and cooperation in research and development, but in particular the communiqué provided for an emergency oil allocation scheme in the event of future shortages and the establishment of a working group to coordinate these measures and submit a plan for a comprehensive action program. Although this working group was also given the task of preparing for a future producer-consumer conference and of carrying out its work through existing organizations, development of forceful consumer cooperation clearly had priority. The communiqué proved unpalatable to the French, who refused to sign important parts of it and criticized the conference for representing a rich group of industrial nations, alienating Arab oil producers, and providing for the reassertion of American domination of Europe. Speaking on French radio shortly after the conference, Jobert charged that Schmidt had chosen the United States over Europe, and that France remained as the one truly European country — a theme echoed by the Gaullist party's Secretary-General, Alexandre Sanguinetti, who accused the U.S. of backing the domination of the oil companies.[67]

France thus seemed to have been separated from her EC partners over the issue of cooperation with the United States. The Eight had found compelling the arguments for a broadly-based oil consumer grouping for which the Community itself provided too narrow a membership as well as U.S. willingness — as the world's largest or second largest oil producer — to arrange for an emergency oil sharing scheme. Of comparable importance were security issues. Kissinger and President Nixon explicitly sought to link agreement on the oil issue to broader problems of security and maintenance of the American commitment to Europe. In Nixon's words to the Conference, "Security and economic considerations are inevitably linked and energy cannot be separated from either."[68] This linkage strategy succeeded in separating the British from their previous alignment with France and bringing them more closely into support of the U.S. position — a change that was solidified by the unexpected electoral defeat of Heath's Conservative government on February 28, 1974, and its replacement by the more Atlanticist (and less European-oriented) Labour government of Harold Wilson. Indeed, the security commitment remained of importance even to France: at the height of the crisis the previous November, Jobert had called for the maintenance of U.S. troops in Europe;[69] but the government of Pompidou, Messmer, and Jobert had both internal political reasons and the lengthy precedent of Gaullist foreign policy to provide a basis for resisting American pressures for linkage between military security and energy policy.

The French effort to galvanize the Nine into a common policy of dealing directly with the Arabs produced yet another temporary policy oscillation by the Europeans. A mere two and a half weeks after the Washington Conference had appeared to confirm the acceptance of the American option — by all except France — the EC Council of Ministers adopted a plan for long-term economic, technical, and cultural cooperation between the Community and a group of 20 Arab states. The Council sought to establish joint EC-Arab working groups to discuss mutually beneficial measures in the fields of industry, technology, agriculture, transportation, energy, raw materials, and financial cooperation. Walter Scheel, in his dual capacity of President of the Council of Ministers and as the person presiding over European political cooperation was authorized to open talks with the Arabs.[70] The reaction of Kissinger and the U.S. government was one of great anger, both at the substantive proposal, which appeared to counter the Washington Conference decision on a joint Atlantic-Japanese approach, and at the failure of the Europeans to consult with the U.S. beforehand.[71] (Though, to be sure, the pattern of non-consultation was a long and mutual one, marked on the U.S. side by such notable occasions as the August 1971 financial measures and the October 1973 U.S. military alert.)

The French government's view of Europe's position was that more extensive steps to promote European unity could be taken only if they served policies which made Europe independent of the U.S. In the words of Prime Minister Pierre Messmer:

> The extension of European cooperation we desire makes sense only if it leads to the affirmation of an authentic European personality, independent of her world partners. That is the fundamental debate, as the Washington Conference showed.[72]

Messmer went on to a sanctimonious analysis of the response to France's pro-Arab policy, and one which also rationalized the large number of bilateral dealings in which France was then engaged: "I make no secret of the fact that the Arab countries were grateful to us for adopting in the Middle East conflict the standpoint of wisdom and of justice."[73] The implication of this message was that France's EC partners must choose between an Atlantic-Japanese response to the energy crisis — which meant French refusal to allow a common policy and further institutional development, or a policy of direct European-Arab dealings — which implied tangible advances in European cooperation.

This French conception was not without support: President Ortoli, for example, emphasized the lengthy ties between Europe and the Arabs, telling an audience of Americans in Brussels that "Europe has particular obligations and interests of her own which may sometimes differ from American interests . . . Europe's dependence on Arab oil (90% of Europe's supply as against only 10% for the United States) is a case in point."[74] As early as January 23rd, the Commission had produced proposals for enhanced European-Arab interchange, suggesting that in return for EC industrial cooperation the oil producers should cooperate by providing supplies at reasonable prices and avoid creating monetary upheaval. The Commission had also requested authorization from the Council of Ministers to express to the producers the Community's readiness to negotiate economic, technical, industrial, and commercial agreements.[75]

Indeed, throughout the crisis and the Washington Conference, the European Community had maintained the position that producer-consumer cooperation was essential and that a confrontation between consumers and producers must be avoided. The lop-sided dependence of Europe on Arab oil explained much of the basis for the Nine's agreement here, but France's eight EC partners were caught in a painful dilemma. To follow the French position meant a serious breach with the United States, which the Germans and then the British found intolerable; to support the American initiatives, however, was to lose the French and thereby any hope of a Common European policy. In the end, given America's energy resources, its economic strength (particularly its limited vulnerability to international resource and financial problems), and its superpower military-political standing, the Atlantic approach seemed to offer payoffs in dealing with tangible problems which the French-led policy simply could not deliver.

For the Eight, this combination of political priorities and efficacy tipped the balance in the American direction, even though the European dilemma continued to cause them to pay lip service to both options. In view of explicit U.S. pressure, the inability or unwillingness of the Arabs to deliver visible short-term benefits, and the accession of more Atlanticist and less European governments in Britain (March), Germany (April), and France (May), the EC-Arab effort stagnated. While the Commission continued to advocate the improvement of relations between the Community and the oil-exporting countries, referring to a "very far-reaching complementarity of interests" between the Nine and the major exporters and asserting that a "climate of mutual trust between the Community and its suppliers" was the best guarantee of stable oil supplies,[76] actual negotiations between the two sides became delayed — ostensibly over the issue of Palestinian representation which took nearly a year to resolve.[77] In any case, had the Nine really been intent upon a Euro-Arab dialogue, the appropriate partner would have been not the twenty-member Arab League but the smaller and more select OAPEC group of Arab oil-producing states;[78] and when the dialogue finally did begin, it remained confined to largely peripheral or non-controversial — i.e., non-oil — issues.

The responses of the Nine to the Commission's series of energy proposals, to the Washington Energy Conference, and to the Euro-Arab dialogue had thus established a pattern which would vary only marginally as new issues arose. That is, the Community's own energy policy would make only halting and insubstantial progress; the Eight would differ from France by deferring to an American lead on crucial facets of international energy negotiations; and some attention would be paid to the desires of France and the Commission to establish rapport with oil-producing countries but mainly to the extent that this did not run afoul of Atlantic priorities.

In September 1974 the EC Council of Ministers finally adopted energy guidelines proposed by the Commission which called for decreased oil imports, increased efficiency of energy use, a coordinated plan to deal with shortages, and efforts to encourage research and development (especially in the nuclear energy field). However, the indecisiveness of this common policy was evident in the fact that the foreign ministers deleted a Commission provision that the Nine should "speak as a Community with other countries on energy."[79] Subsequent efforts largely provided reformulation of these guidelines, such as plans submitted in December 1974 to reduce the growth rate of energy consumption from a projected 5% per year to 3.5% so that by 1985 the share of oil in Europe's energy balance would have declined from 61% to 49%.

Essentially, energy policy differences among the Nine remained unresolved and subject to the strong pull of domestic political priorities. For example, in late July 1974 the newly elected British Labour government followed the example of its Conservative predecessor by blocking further progress on a common energy policy. Whereas the Heath government had done so in order to bargain for a larger, and domestically important, regional policy, the Wilson government feared stringent control of energy policy from Brussels. In response, a proposal for a Community energy agency with its own revenues from tariffs on imported energy to finance research and exploration was abandoned. Yet the British persisted for nearly three months in blocking an EC energy policy aimed at scarcely more than reducing Europe's reliance on energy imports.[80]

Because of domestic considerations even Germany raised obstacles to EC measures in dealing with the energy crisis. In this case, the new government of Helmut Schmidt, seeking to rebuild Social Democratic popularity, insisted on stringent conditions and limits upon EC acceptance of up to $3 billion in medium term loans from Arab oil producers to the central banks of member states in balance of payments difficulties. Since Bonn would be expected to provide security for each loan up to a maximum of 44%, the German government sought to avoid appearing to its electorate as the milk cow for the Community unless tangible links to European progress could be demonstrated.[81]

It was thus no wonder that the Community's Commissioner for Energy could lament that "there is more Atlantic cooperation than European unity in energy."[82]

Not only did success in delineating a common energy policy within the Community and in dealings with oil-producing countries prove limited, but efforts to maintain EC solidarity in negotiations in the Atlantic context were also restricted. Even after the Washington Energy Conference in February and the election of Valérie Giscard d'Estaing to the French presidency in May 1974, France remained divided from the Eight over priorities. While the French did not obstruct the operations of the Energy Coordinating Group (which resulted from the Washington Conference) — as they were capable of doing through the institutions of the European Community and the OECD — they chose to remain outside the newly created International Energy Agency (IEA) which was likewise an offshoot of the conference. For a time, after the establishment of the IEA in November and the Martinique summit meeting of Presidents Ford and Giscard d'Estaing in December 1974 it appeared conceivable that France might join the IEA after all, thus presenting a united consumer front on energy as well as creating the possibility of the Nine acting as a bloc within the IEA. Indeed, as Finance Minister, Giscard had wanted to attend the Washington Energy Conference and had been prevented from doing so by President Pompidou. Yet despite support for the idea at the Quai d'Orsay, French domestic politics intruded. The Gaullists, making up the majority of Giscard's parliamentary majority, together with the bulk of the Socialist-Communist opposition, vociferously opposed such a concession to American hegemony. Indeed, among both Gaullist and opposition ranks, there existed a commonly voiced — but unsubstantiated — assumption that the Americans had provoked the energy crisis in order to reimpose their economic and political domination of Europe. Apart from this belief, however, it was widely observed that the U.S. had benefited from the crisis both economically (through her multinational oil companies and the weakening of rival economies) and politically (by the reassertion of her leadership). In the words of the French Socialist leader, François Mitterrand, addressing a Socialist International Conference in West Berlin: "The U.S. profits from the situation in order to reinforce their economic domination over Western Europe."[83] France thus remained outside the IEA, although privy — via the Community — to prior and non-binding consultation with the Eight on the substance of their policies within the Agency.

27

The French government's position of benign neglect, rather than overt antagonism, toward the IEA was also shaped by the desire to preside over a preparatory trilateral meeting at which France would lead in mediating among the principal industrial consumers (the U.S., Japan, and the EC), the oil producers (represented by Algeria, Saudi Arabia, Venezuela, and Iran), and the less developed consumers (India, Zaire, Brazil). Once again, differences between the Eight and France proved unbridgeable. The Foreign Ministers of the Nine failed to agree on a common role to play in the preparation of the trilateral conference. Whereas the Commission had sought such a conference as well as proposed that the Eight not subscribe to any decision within the IEA without first deliberating à la Nine (i.e. in the presence of France), several member states — the Netherlands, Denmark, Germany, and especially Britain — made it clear that they preferred to work with the U.S. and that they would not limit their freedom within the IEA, or at the preparatory conference on behalf of the Community.[84]

France, as a non-member of both OPEC and the IEA, chaired the Paris preparatory meeting of the trilateral energy conference. In the IEA and at the conference the Eight had agreed with their French partner to be bound by a "common position" when it existed;[85] otherwise they were to be free to act as their national interest dictated. As it happened, the April 1975 preparatory conference collapsed in disagreement. Although France had managed to win American and EC support for holding the meeting, the differences among the developed consumers remained unresolved. Whereas the oil exporters and the poorer consumers demanded a wide agenda to cover stabilization of other commodities, transfer of technology, monetary reform, and investment, the industrialized consumers (particularly the U.S., Germany, and the U.K.) preferred to concentrate exclusively on energy matters. The French were unable to bridge this gap and the talks concluded without reaching agreement on an agenda for the putative trilateral conference.[86]

In sum, throughout more than a year-and-a-half of the energy crisis, the European Community and its members had made little progress in formulating either their own common energy policy or in pursuing a common policy in international negotiations and organizations. There is no reason to disagree with the assessment of the Commission President, François-Xavier Ortoli, that the crisis left the U.S. more dominant and the Community weakened in its influence on issues of security, finance, and economics because of its lack of a single voice.[87] The one tangible Community achievement — accomplished with the help of the members of the OECD and of the Group of 10 and Committee of 20 of the IMF — had been to prevent the onset of a situation

28

like that of the international economic crisis of the 1930s, replete with import restrictions, competitive currency devaluations, autarchy, and siege economies. Yet even here, coordination in meeting these problems, for example, in stabilizing Italy's financial difficulties, owed as much or more to multilateral coordination (e.g., in the IMF, OECD, IEA) as it did to the Community's own action. Of the three options available to the Nine and pursued by them — bilateral, multilateral, and Community — their response as a group was by no means the most successful. As we shall see in the next two sections, the members of the EC had courses of action available outside the Community which they undertook individually and in groups.

b. National and Bilateral Responses

The various Community and multilateral efforts were accompanied by specific national responses to the energy crisis from each of the nine EC governments. While the events of the winter of 1973-74 may have made the Europeans increasingly aware of their shared predicament and identity, the experience also underlined their vulnerability and their inability to formulate and maintain a coherent regional response. In these circumstances, which stimulated faint reminiscences of wartime shortages and deprivations as well as some real concern about being starved of the petroleum needed to maintain their industrial economies and indeed assure the survival of their societies, national governments undertook measures to safeguard foreign oil supplies. Some, particularly Germany and the Netherlands, believed that national interests would best be protected by a coordinated multilateral policy among the principal oil-consuming countries, not least the United States. Others, especially Italy, Britain, and France, placed greater emphasis on striking bilateral deals with oil-producing states, initially to guarantee security of supply and subsequently to provide for means of monetary recycling to cope with the 400% increase in oil import costs.

While the French may not have invented the term *sauve qui peut* ("every man for himself"), they were widely perceived to be its most successful practitioners. Ironically, there were rumors that the French government may have favored a hard-line joint consumer response to break the oil cartel in the first month or two of the crisis, or at least to provide unified resistance to price increases. Although NATO Secretary-General Joseph Luns spoke of the Arab oil embargo as tantamount to a declaration of war, he was isolated even by the American government, and in the absence of a firm Western response, the French found it convenient to go their own way as expeditiously as possible.[88] They also saw an opportunity to regain influence in the Middle East lost to the British and American governments and the major oil companies of those countries during the aftermath of the First and Second World Wars.

By the end of January 1974 the French had sent emissaries to the principal Middle Eastern oil-producing states and concluded provisional negotiations to buy 800 million tons of Saudi Arabian crude oil over a period of 20 years, an amount estimated to cover one-fourth of France's oil consumption needs. At the same time the French agreed to sell Saudi Arabia Mirage fighter planes and other heavy weapons.[89] Almost simultaneously it became known that France and Kuwait were negotiating another weapons-for-oil bargain. In both cases an additional *quid pro quo*, which the Saudis and Kuwaitis proposed to Foreign Minister Michel Jobert, was that the oil producers expected France's cooperation in blocking establishment of an oil consumer group.[90] Jobert's denunciation of the February Washington Energy Conference as a "provocation" logically followed from France's strategy of bilateralism.

While France pressed on in negotiations with Libya, Iraq, Algeria, Syria, and the (non-Arab) Iranians, her EEC partners began to intensify their own efforts. In addition to approaches by Italy and Belgium, Britain sought to maintain or establish special relationships with her erstwhile protectorates among the Persian Gulf oil sheikdoms as well as with Saudi Arabia and Iran. During the initial stages of the embargo the British had shared membership with the French in the Arabs' most-favored category, exempt from oil supply reduction. King Faisal of Saudi Arabia, seeking to encourage the pro-Arab policy of Britain's Conservative government and hostile to what he regarded as the leftism and pro-Israeli position of the opposition Labour Party, actually ordered that the UK be given more Saudi oil than before the boycott.[91]

Prior to the crisis the EC Commission had sought to discourage competitive bidding for oil supplies among the Nine and between them and other major oil importers. However, by late January 1974 it could do no more than observe that the problem of achieving stable energy supplies at reasonable prices "could not be solved by a series of bilateral agreements along the lines of barter deals,"[92] as opposed to overall cooperation between the Community and oil producers. Indeed, the Commission's response to the lack of agreement on energy policy and the virtual stampede by EC members to bid for bilateral agreements and special treatment at the hands of the producers embodied an element of pathos:

> Until the Council has handed down a decision, the agreement which might be concluded by certain Member States with the oil-

30

producing countries should at the least be subject to previous consultation at Community level, as the Commission has suggested.[93]

Translation: Kindly don't scramble all over each other in your rush for the exits, but if you are going to do it anyway, at least be good enough to tell us about your deals.

The Commission was not the only actor aghast at the display of *sauve qui peut*. Finance Minister Helmut Schmidt expressed the West German government's dismay at the "go it alone trend" of the British and French, and threatened that this would cause Bonn to hesitate on her commitment to the EC Regional Fund.[94] And the United States trade envoy, Harold Malmgren, bitterly denounced a British-Iranian deal (5 million tons of crude oil in exchange for $240 million worth of textile fibers, steel, paper, and other industrial goods) as "aggressive bilateralism," "naive and dangerous."[95]

In effect, a struggle had begun to develop between the U.S. and Germany on the one side, and the oil producer countries on the other; the former sought to rally the developed oil-consuming countries to joint, multilateral action in dealing with energy supply and price problems and their consequences, the latter to deal with the consumers on the basis of individual *bilateral* arrangements. Both sides hoped to attract the loyalties of the governments and bureaucracies, as well as of groups and corporations in consumer countries such as France, Britain, Italy, Belgium, and Japan, who realized that their national interest required secure oil supplies. In this contest, the Community's institutions and policy cooperation were virtually overwhelmed. Although Michel Jobert's successor as French Foreign Minister, Jean Sauvagnargues, sought to play down the divergencies, noting that France's position on energy policy "can be summarized in one formula: reduction of consumption and long-term solidarity among the producer and consumer countries,"[96] this slogan avoided the question of the extent to which the interests of producers and consumers might differ. By contrast, the position of the militant Libyan government was directly aimed at dividing the consumer countries. At a Paris dinner given in his honor, the Libyan Premier, Abdul Salam Jalloud, spoke of the need to create a "real Arab-European economic force" and observed:

Our European friends must accept the fact that if Europe's interests coincided with those of the U.S. in the nineteen forties and nineteen fifties, that has no longer been the case in the nineteen sixties and nineteen seventies.[97]

31

After the European countries recovered from the initial shock and realized that the supply of oil was less likely to be a pressing problem than its price, the pattern of bilateral dealings began to shift in emphasis. The French government, which had signed oil deals with Libya, Kuwait, and Saudi Arabia, and which was in the midst of discussions with Iraq, Iran, and Algeria, began to hesitate on the matter of price. Shortly before his death, President Pompidou in March 1974 told a ministerial meeting dealing with energy that France would not sign any agreements stipulating higher prices than those charged by the international oil companies. The Saudi accord signed during the height of the crisis had been at a price too high for the French domestic market. Although the actual commitment had been for the relatively modest amount of 30 million tons for each of three successive years, the vast remainder of the tentative long-term French-Saudi deal would have involved unacceptable costs for France if consummated. In the case of the Libyan agreement, Giscard d'Estaing (then Finance Minister) actually prevented the French national oil company (ELF-ERAP) from carrying out a deal to buy crude oil at $16 per barrel (a figure above prevailing world prices and as much as $3 per barrel more than what the company could have sold oil in France for without a loss). Finally, an agreement with Algeria for limited quantities of oil also fell through because Algeria's asking price was too high. It was at this point that the French government finally began to impose domestic measures (restricted speed limits and lowered home temperatures) designed to conserve energy, and later a cost ceiling on the 1975 imports of energy at a level 10% below that of 1974.[98]

From the spring of 1974 onward, the pattern of bilateral bargaining involved less bartering or bidding for long-term oil deliveries than it did efforts to cope with the massive foreign exchange costs of imported oil. Attention now turned to finding increased markets for exports of industrial products, armaments, and technology in order to offset growing balance of payments deficits and to encourage the flow of petrodollars from the Middle East back to Europe in the form of loans and investments. Once again France proved most adept, by signing, for example, in June 1974 an elaborate 10-year accord with Iran for some $4 billion worth of contracts and industrial plants covering the sale of five nuclear reactors, development of Iranian natural gas, electrification of railroads, subway construction, petrochemicals, and the training of scientists and technicians.[99] In contrast to agreements concluded only a few months earlier, oil was not directly involved in the deal. By the end of 1974, the French had negotiated a series of agreements with Iraq and Algeria, for $3 billion each, with Saudi Arabia ($800 million), and had increased the value of the Iranian deal to $7 billion.[100] The strategy of these bilateral transactions involved a stress on French industrial exports in order to cope with the impact of oil import costs on the balance of payments and as the preferred means of recycling petrodollars, rather than by encouraging the oil producers to invest in France. The French policy, neatly summarized by Finance Minister Jean-Pierre Fourcade at a September 1974 IMF meeting, was that OPEC would never be allowed "to buy up half of French industry so that the other half can have the oil it needs to keep going."[101]

To a substantial degree this policy was also dictated by French domestic constraints in terms of the need to maintain full employment and to cope with France's major long-term disadvantage — similar to that of Japan and Italy — in dealing with the energy crisis. Unlike the U.S., Britain, or Norway, France did not have substantial energy resources of its own, and in contrast to Germany it lacked an unusually resilient industrial base and a long-term balance of payments surplus. In consequence, the government also took major steps to develop a vast national nuclear power program to provide 23% of French energy needs by 1985 and 45% by the year 2000.[102] Once again the French followed a bilateral policy rather than a European one, in this case selling Iran a share in the French Eurodif uranium enrichment plant after being unable to get the agreement of Britain, Germany, and the Netherlands, all of whom preferred a different, cheaper, centrifugal process.[103] Even French participation in the European Community could be put to the service of bilateral dealings and French domestic interests. Thus, as a follow-up to a commitment made in negotiations with the Iranians, French Foreign Minister Sauvagnargues even proposed that the Community negotiate a preferential trade agreement with Iran, a proposal which irritated the Commission and the Ford Administration because its enactment would have violated Commission assurances (by Vice-President Christopher Soames to U.S. State Department officials in 1973), that the Community would confine its preferential agreements to Britain's former EFTA partners, the Mediterranean, and a group of developing countries (mostly former French and British colonies).[104]

In the pattern of French response, whether strictly bilateral, or in making use of the Community, or even in remaining on the threshold of the International Energy Agency while actively involved in the IMF and OECD, there is a common thread. It is a strategy emanating from a certain definition of French domestic and international interests. Since the impact of transnational actors and forces (oil companies, OPEC and OAPEC, monetary and trade flows, the IMF, the EC, the IEA) could produce severe dislocations in the economy, society, and politics of France, the French government, in seeking to cope with these changes

33

and to maintain internal equilibrium, would tend to pursue a policy in which membership or obligations in any external organizations or groups would only be given weight when they provided a means for coping with these problems. If circumstances dictated aggressive bilateralism, or Community solidarity, or even multi-lateral consumer cooperation, the appropriate course of action would be taken and then sanctified with some elastic but elegant rationale. Ironically, this freedom of maneuver was made possible by the continuation of an East-West balance of power in Europe, based on the maintenance of an American troop commitment to the defense of Western Europe. Thus, at the height of the crisis, in a speech highly critical of the United States, Michel Jobert could call for the continued presence of American troops even while continuing to resist any linkage to other non-military issues as a *quid pro quo* for this commitment.[105]

The domestic success of the French government in carrying out these policies was evident in public opinion figures. A February 1974 SOFRES poll showed that by a 52% to 28% margin, the French public approved of its government's decision to refuse to follow the U.S. energy proposals even though the Eight accepted this line. Majorities or pluralities among Gaullist, Communist, Socialist, and even Reformist (centrist) voters all expressed approval.[106]

With changes in the governments of Britain (March 1974, from Conservative to Labour) and Germany (April, from Brandt to Schmidt), France's principal EC partners were less inclined to bilateralism and more favorably disposed toward multilateral cooperation with the U.S. Nonetheless, even these administrations sought to cope with the monetary impact of the energy crisis by negotiating bilateral export or investment arrangements. Britain, for example, obtained a $1.2 billion line of credit from Iran in July 1974 and in January 1975 signed a trade protocol for a similar amount of export business with the Iranians. In pursuit of a nationally beneficial regional policy the British had held up the Community's energy policy for a time and later began to pursue a North Sea oil policy which sought to maximize national control of this vital resource and minimize EC involvement in it. Germany allowed Iran to take a 25% interest in her huge Krupp steel operation, and Italy also reached long-term agreements with Iran with a potential $5 billion in joint projects. Even the United States, despite Secretary Kissinger's stress on the need for the consumers to act together rather than seek individual oil producers,[107] concluded a 5-year trade pact with Iran worth $15 billion ($12 billion of which would be spent in the U.S.).[108]

It is certainly possible to overestimate the extent of *sauve qui peut* reactions. As a high EC official has observed, most countries (including the U.S.) had had a pre-crisis pattern of various dealings with oil-producing states and during the crisis itself few of the many deals negotiated ultimately involved real oil commitments. Bilateral economic agreements as a means to recycle petrodollars did not involve the obnoxious bartering and competitive bidding for oil which the Community and the U.S. had sought to discourage, and indeed one result of the scramble that did occur may have been to discredit the utility of such responses in the event of a future crisis.[109] Certainly there were only limited payoffs for those who engaged in aggressive bilateral bargaining. While France did achieve enhanced export markets and perhaps a slightly greater degree of influence within the Arab world, her oil companies nonetheless suffered prorated supply cuts roughly comparable to the shortfalls of other consumers, even those with less favored embargo status during the October 1973 - March 1974 period. This was due to the fact that in the absence of any other effective machinery the major international oil companies themselves allocated world oil supplies. And, for all the accords and agreements negotiated, France did not obtain any oil price advantages over other consumers. Indeed, France even gained very little political influence. French officials sought to achieve a role for France as *interlocuteur valable* between Arabs and Israelis and between producers and consumers. But although Foreign Minister Sauvagnargues told the Arab League and Yassir Arafat that France could not envisage a Palestinian solution which would lead to the destruction of Israel,[110] France could deliver no bargain with the Israelis because of what the latter perceived as a seven-year history of one-sidedly pro-Arab policies. Even among consumers and producers, France could not quite bring off a successful set of negotiations. While Giscard did succeed in convening the preparatory Paris Conference in April 1975, he lacked the ability to win over major consumer countries inside and outside the EC to his point of view. Since France could thus deliver neither a Middle East settlement nor a producer-consumer agreement, the political influence she could exercise through her policies of bilateralism remained seriously limited.

In general, Machiavellian policies of bilateralism and narrowly national responses seemed unlikely to offer broadly successful courses of action for individual EC countries, whether their concern was with security of supply, oil prices, or even political influence. Yet this did not mean that certain individual agreements, particularly those involving exports of technology, industrial goods, joint ventures and the like were not useful as a means of recycling the flood of petrodollars.

To the extent that measures of bilateralism were pursued during the crisis and might again be in the future, the causes would lie as much in domestic political and economic necessities as in the failure of the European Community and broader multilateral organizations to offer a more promising means of joint cooperation. In a sense, the issue of oil, particularly if the chief aspect appeared to be security of supply, would be more likely to generate a *sauve qui peut* response than other kinds of economic and resource problems. On the one hand, the absence of agreed and effective mechanisms of consumer cooperation left each state vulnerable to the logic of Prisoner's Dilemma, each fearing that if, out of a sense of loyalty, it hesitated to cheat by making a special deal of its own, then one of its neighbors would have a strong motivation to jump at the opportunity first.[111] Here, too, there existed significant domestic motivations for such behavior. Since a government may be expected, above all, to protect national security by appropriate means, strong domestic pressures for safeguarding vital energy supplies, whatever the means, will emerge once the supply of oil becomes mainly a matter of politics rather than of economics and technology.

While this *sauve qui peut* logic can also be linked to balance of payments problems — as indeed proved to be the case during the catastrophic economic crises of the 1930s, there do exist mechanisms in this sphere for the mutually advantageous management of monetary policy without resort to measures of competitive devaluation, autarchy, and the like. Thus, as the energy problem moved from an initial concern with supply to monetary-economic considerations, the chief alternative to a coherent EC response shifted from nationally based policies of bilateralism toward multilateral cooperation around organizations such as the OECD, the IMF, and the IEA, all of which transcended the EC context.

c. *Multilateral Responses*

Although the Europeans found it necessary to employ multilateral measures in dealing with the energy crisis, this did not automatically result in a response at the Community level so much as in cooperation among the major non-Communist industrialized consumer countries, i.e., among the U.S., Japan, Canada, and the European non-members of the EC, all of whom participated with the Nine in the OECD.

The principal reason for this broader-based cooperation was the fact that the U.S. possessed resources the others, particularly the Nine, lacked. First, the Americans were not only the world's leading oil producer but also held major energy resources of coal, shale, and nuclear power. Whereas the Nine imported 63% of their overall energy requirements (and 98% of their oil), the U.S. imported only 17% of its needs (and 38% of its oil).[112] Second, the U.S. continued to possess a broad military and political leadership which — in the absence of real European political unity — made it the logical leader in any coordinated Western consumer response. Third, and above all by virtue of its size, resources, and lesser dependence on international trade and payments, the U.S. enjoyed a measure of economic and monetary strength which left it far less vulnerable than the Europeans to disruption of the international economic system. This fact had been somewhat obscured by the monetary problems and dollar devaluations of the 1971-73 period, as well as by the accompanying chorus of voices on both sides of the Atlantic announcing the demise of American economic leadership. The energy crisis, however, once again brought into sharp relief the picture of economic strength based on near self-sufficiency, continental size, and a gross national product greater than that of all the Nine combined.

The implications of this were not yet felt by the Europeans in the initial stage of the crisis (approximately, the weeks from mid-October to late December 1973). In this period the climate of European-American rivalry which had antedated the crisis and characterized economic, monetary, trading, and agricultural import issues, spilled over into initial reactions to the October War and the threatened energy shortage. There were not only sharp differences with the U.S. over the conflict in the Middle East, manifest, for example, in the restriction of military bases and supply facilities for U.S. use in the resupply of Israel, but it also appeared conceivable in the immediate aftermath of the war that the Nine might achieve a modus vivendi with the Arabs over technical, economic, energy, and even certain political questions. These arrangements would be reached independently of the U.S., reflecting not only Europe's enhanced status, but also an underlying difference of interest, geography, and energy resources — and America would be left to pull its own economic and military chestnuts out of the Middle East fire.

European leaders entertaining these notions were soon disabused of them. The limits on both Community and bilateral responses to the crisis, together with the re-establishment of fundamental American strength in the economic, energy, and security spheres, rapidly made themselves felt. Over the next few months the pattern of multilateral, primarily Atlantic, responses became increasingly important.

At least as early as 1972 the U.S. had expressed official readiness within the OECD to undertake consumer cooperation in order to increase the availability of energy supplies, decrease the reliance on Middle East oil, and provide for a coordinated response in the event of supply restrictions.[113] The American proposals were, however, insufficiently attractive to the Europeans because they were based on the sharing of imported oil alone rather than on overall oil consumption. Ultimately, the

Washington Energy Conference of February 1974 proved a turning point in the formation of a multilateral response. Over French opposition, the Eight, together with Canada, the U.S., and Norway established an Energy Coordinating Group (ECG). This body in turn drafted an International Energy Program (IEP) which provided the basis for the establishment of the International Energy Agency (IEA) in November 1974.

Essentially, the IEA offered a detailed plan to cope with oil supply problems in the event of a future embargo and a commitment to establish 90-day reserve stocks of oil. By bringing together countries responsible for four-fifths of world oil consumption it also provided a means of consumer solidarity in dealing with producers. Finally, it imposed commitments to conserve energy in order to reduce imports and to encourage research efforts toward the development of alternative energy supplies.

Although initially composed of twelve countries, IEA membership was open to all members of OECD (within whose framework the agency was established, but with its own governing board and secretariat). In rapid succession, the U.S., Japan, Canada, and the Eight (Norway having reduced its status to associate membership) were joined by Austria, Spain, Sweden, Switzerland, Turkey, and New Zealand. France remained outside the IEA but in effect closely associated by virtue of its eight EC partners holding full membership in the Agency. Despite periodic French criticisms of the IEA because it was allegedly in potential conflict with the Community's energy policy and also involved risks of "confrontation" as opposed to France's policy of "concertation" between producers and consumers,[114] France made no effort within the OECD or the EC to obstruct the Agency's establishment. She thus stood to gain many of the benefits of membership while maintaining freedom of action.

At the heart of the IEA lay its quasi-automatic oil-sharing formula involving a "strong presumption" of action in the event of a crisis.[115] If one or more IEA members were to suffer supply cuts amounting to 7% or more of their previous year's oil consumption, then all would be obliged to reduce their own oil consumption by 7% and to cooperate in the reallocation of available oil imports. If the consumption loss were to amount to 12%, then there was to be a 10% reduction in consumption in addition to some use of reserve oil stocks. Losses greater than 12% were to lead to agreed measures of further demand restraint. The novelty of the arrangement lay in its automaticity. Once the 7% shortfall (lesser amounts having represented a kind of self-insurance or "deductible") was reached, the sharing formula would come into operation unless

38

blocked by a weighted (60%) majority vote of IEA members.[116] Neither the U.S. nor the Eight could block such a majority without the support of at least one other country.

In calculating oil shortfalls, not only oil imports, but stocks and domestic production were included, thus surmounting a previous European objection to U.S. sharing proposals. Whether in some future crisis the U.S. would actually export scarce supplies of its own oil to the Europeans was another matter. In the view of a high IEA official, it was doubtful that such a situation would ever arise; instead, the emergency oil sharing formula was meant to insure that all parties "go to the wall" at the same rate.[117] This was by no means a counsel of despair. By one calculation, the IEA members could endure an Arab embargo twice as severe as that of October 1973-March 1974 (which at its peak curtailed oil supplies by 10%) and do so for a period of 600 days by cutting their consumption by 10% and meeting 10% of their needs out of existing 60-day stockpiles.[118] According to a subsequent calculation by the IEA head, Etienne Davignon, Western Europe could endure an embargo for 400 days by reducing its consumption by 15%.[119]

Unable to achieve agreement on a significant EC energy policy, the Commission sought to preserve at least a single Community voice on energy within the IEA. In October 1974, the Commission formally approved the membership of the Eight in the IEA, but Henri Simonet, the energy commissioner, warned that they would be in violation of Common Market rules if they participated in crisis oil sharing which violated the free movement of goods "enshrined" in the Treaty of Rome. Since France was the only Community member outside the IEA, the Eight were in effect told that France would enjoy the benefits of emergency oil sharing without having had to enter into any direct commitments.[120] In December, the Commission achieved observer status within the IEA, and in January 1975 the EC Council of Ministers accepted a Commission suggestion that, in view of Article 116 of the Rome Treaty requiring that in economic relations with international organizations the EEC must act as a single entity, any important issue involving the Eight within the IEA would be harmonized among the Nine in advance.[121]

Apart from its emergency oil-sharing formulas and the issue of EC participation, the IEA-OECD grouping offered the Europeans a broader base for coping with some of the financial implications of the energy crisis. In April 1975, responding to an earlier proposal by Secretary Kissinger, the 24 OECD countries signed a $25 billion agreement providing for a loan facility and financial safety net to aid members in difficulty as a result of energy import costs. The group also remained committed to the principle of a dialogue on economic cooperation between producers and consumers, though, as the OECD delicately phrased it, "Such cooperation is likely to be more effective if the OECD countries can achieve an effective cohesion in the field of energy."[122]

The primacy of a multilateral response to the energy crisis is sometimes held to be the inescapable result of growing economic interdependence and the impact of transnational forces and actors. In the view of Henry Kissinger, joint action by North America, Western Europe, and Japan was necessary to avoid perpetual crisis and to protect the Western world from both the threat of a new embargo and the risk of financial collapse.[123] Such a collapse in Italy or Britain, for instance, would affect everyone, therefore broad cooperation must be established to prevent such dangers. From this perspective, the European Community was too narrow a grouping, whereas the OECD provided a broader and more flexible format for mutual action by the developed industrial states of the non-Communist world. Not surprisingly, such views tended to be held by high officials of the IEA, but elsewhere, too, the establishment of the agency was characterized as the most fruitful U.S.-European system initiative since the Marshall Plan and the North Atlantic Alliance.[124]

In practical terms there are, however, both real and potential obstacles to any unqualified estimate of a wholly successful IEA. A fundamental test would come in the event of a future oil embargo. IEA officials believed that their emergency oil-sharing formula primarily provided a deterrent making a renewed embargo less likely, but they expected that if an embargo did recur their members could at least cope with a supply problem of moderate duration and intensity. They also pointed to the likelihood that lessons had been learned from the initial crisis, namely that policies of bilateral *sauve qui peut* did not offer an effective alternative. In addition, the availability of cooperative machinery, absent during the winter of 1973-74, was likely to further lessen the probability of a panic response. IEA members would also tend to hesitate before violating solemn international commitments to the agency and their co-members.

Estimates of the IEA's viability in a crisis and of whether the Nine would respond on a multilateral rather than a Community or bilateral basis differed among officials of the European Community. Some shared the expectation of a successful cooperative IEA performance. Others anticipated a European disarray comparable to that of 1973-74, in part because of Europe's continued energy vulnerability, in part because of the absence of U.S.-European agreement on policy toward the Middle East, and in part because of a basic divergence of U.S.-European interests. In this pessimistic view, much of the European response would depend on how a new Middle East conflict erupted (i.e., who did what to whom first). This view also held that if a selective embargo were imposed there might be a strong temptation — once again — for the non-embargoed to steer clear of any involvement in aiding their partners. Although OAPEC members might be less inclined to impose an embargo because of its limited prospects of success and the risks of alienating the Europeans as well as provoking a possible military response by the United States, the dynamics of Arab domestic opinion and of pressure from the more bellicose Arab states and actors might very well precipitate another embargo in any case. Even if the oil-sharing formula did come into operation, it was by no means clear whether the multinational oil companies would or could carry out the allocation as directed by the IEA and its member governments. In 1973-74 the companies had handled oil allocation on a pro rata basis, violated long-term contracts in order to do this, and resisted pressures from individual European governments seeking unilateral advantages. The companies remained a chief source of information on oil production, trade, shipping, pricing, and the like. While Article Six of the International Energy Program provided that the IEA governing board would decide on practical procedures for allocation and for the role of oil companies, no one could be certain of the future response of a major international oil company caught between conflicting demands of a host, producer government, and one or more consumer governments — or even between differing instructions from two consumer governments.

Even apart from the primary question of emergency oil sharing or the deterrence of an embargo, there existed potential differences of interest among IEA members. In order to encourage the development of new energy supplies and technologies, the United States government proposed a "floor" price for oil so that OPEC would not be in a position to destroy these high-cost alternatives by suddenly dropping the price of oil below the figure needed to maintain their commercial viability. As a possessor of vast coal, oil, oil shale, nuclear, and geothermal resources, the United States (and American-based international oil companies) had a logical interest in a high floor price. But the Nine, apart from Britain, had no domestic oil production of their own and therefore favored the lowest possible international oil and energy prices — as did many domestic political critics within the United States.

In sum, while multilateral solutions were mostly employed to cope with the monetary consequences of the energy crisis, to sustain interdependence, and to prepare for handling a future crisis, the bilateral and Community-level responses were by no means foreclosed for the Europeans.

TABLE 4

Energy Policy Integration

	Bilateral	European Community	Multilateral (IEA/OECD)
Institutions	7	6	5
Social Bases	5.2	4.3	4.5
a. Transactions	(4)	(4.5)	(4)
b. Elite opinion	(5.5)	(4)	(4.5)
c. Mass opinion	(6)	(4.5)	(5)
Policy Cooperation	6	6	4.5
Average	6.1	5.4	4.7

The relative importance of the three principal modes of response for the Nine in dealing with energy can be suggested in terms of the degree of integration to be found in each of them. Thus, for the bilateral, European Community, and multilateral responses it is possible to consider the extent of integration in terms of institutions, social bases, and policy cooperation.[125] The category of social bases can also be subdivided here into transactions (e.g. trade, communications, etc.), elite opinion, and mass public opinion. In turn, the extent of integration of energy matters (or of support for it) can be assessed by an adaptation of Leon Lindberg's 7-point scale, ranging from "1" (where matters are entirely centralized or integrated) to "7" (where energy is dealt with on an exclusively national basis.)[126]

An examination of Table 4, where numerical values have been assigned to each of the categories,[127] indicates that in an *institutional* sense (common institutions and jurisdictions) no integration of energy matters exists for bilateral dealings, but that the EC has provided a slight institutional basis and the IEA a somewhat larger foundation. As for *social bases*, the high dependence of Europe on Middle East oil and trade as well as its integration into the Atlantic trade and payments framework, make the degree of bilateral and multilateral integration comparable. Within the EC this integration is slightly less. However, European elite opinion is probably a little more inclined to multilateral and especially European approaches to energy than to bilateral ones. In turn, mass opinion is perhaps less favorable to the bilateral orientation and nearly as favorable as elite opinion to both European and OECD-type responses. Finally, the actual degree of *policy cooperation* (the ability to act as a group in making policy decisions) has proved to be minor and roughly comparable at bilateral and EC levels, and substantial on a multilateral basis. The average scores over all three categories are therefore: bilateral, 6.1; EC, 5.4; and multilateral, 4.7. In short, the figures suggest a rank ordering of the relative importance and degree of integration to be found in each of these areas.

3 FAILURE OF THE EC ENERGY POLICY: SOME BROADER IMPLICATIONS

Contrary to some expectations that the October War and the resultant energy crisis might bring the Europeans closer together, the pressures of these events produced the opposite effect, exposing the Community's "inability to face a major challenge in a way commensurate with its claim to be a major economic power evolving progressively into a political one."[128] As the dependent variable, an ideal European energy policy would have included complete harmony within the EC on taxes, conservation, circulation of supply, and policies toward the oil companies; externally, it would have entailed neither competitive bidding nor bilateralism, but rather a unified approach in dealing with the Arabs and the Americans. In reality, by these measures, there existed very little that could be called a common energy policy.

Why, then, did the Community fail to cohere? Why was its unity undercut by both the bi- and multilateral responses of its members? Why did the demands upon EC institutions overwhelm their capabilities for response? Why, finally, did energy, which had been crucial in the inception of European unity become the catalyst of a profound division? The answers may be found in the way in which these crisis pressures brought to the surface fundamental political differences among the Nine, asymmetries of power between Europe and America, and problems of domestic politics.

a. The Politicization of Energy Policy

Underlying political differences had long existed among the Nine and were likely to impinge on any serious effort to develop a common energy policy. These differences included, first, the question of whether to treat domestic oil companies in the largely laissez-faire manner favored by Germany, the United Kingdom, and the Netherlands, or to pursue a decidedly *dirigiste* or interventionist policy as the French and Italians had done. There was, secondly, the question of policy toward the Israeli-Arab conflict, with France, then Italy, and Britain under the Conservatives willing to adopt increasingly pro-Arab policies in order to safeguard their interests in the Middle East, while the others, particularly the Netherlands, Germany, and the Labour government of Britain were more inclined toward an impartial approach or one which leaned toward the American position. Thirdly, there existed profound differences in orientation toward the United States. The government of France (along with European Communists, some Socialists, and increasingly Gaullist elements among the British Conservative Party) preferred a more autonomous relationship, while the Eight broadly favored close Atlantic ties. In the often quoted words of Henri Simonet, a common policy on oil is "10% oil and 90% politics",[129] and as the energy crisis became rapidly politicized, these and other differences quickly came to the fore.

"Politicization" here is an elusive term but one important enough to warrant defining in order better to understand the dynamics of the European situation. The term is not used to designate the inherent substance of an issue, a procedure which gave rise to the supposed distinction between "high politics" (e.g. security) and "low politics" (e.g. welfare and economic issues), and which has become manifestly inadequate at a time when "economic" issues have become highly political (inflation, unemployment, oil), while security issues are sometimes less urgent. Instead, politicization applies to the overall treatment of a policy issue. I have elsewhere identified three necessary indicators of politicization:[130] 1) the handling of an issue by a primarily political ministry (such as a Foreign Ministry or office of a Prime Minister or Chancellor) rather than by an economic one, the two types of ministry being differentiated on the basis of whether they make limited or extensive provision for group access (involving routinized consultation and bargaining); 2) the existence of involvement by the broader public, which implies attention by at least the attentive public as well as the communications media, thus making treatment of an issue public rather than private; 3) participation by the political parties, which, since public opinion sets only the vaguest limits on foreign policy, interferes with the largely technical consideration of an issue in a closed relationship between pressure groups and administrative departments and also may increase the possibility of policy judgments being rendered on some broader national interest rationale rather than on a narrower calculation.

Joseph S. Nye and Robert O. Keohane have offered a related definition which, while similarily focusing on the treatment rather than the substance of an issue, sees two separate types of politicization: one involving the attention of elite decision-makers to an issue (which may remove it from the realm of transnational and transgovernmental bargaining), and the other coming into play when public attention becomes focused on an issue.[131] In the present case, energy policy became politicized in both senses of the term: the elite attention bringing into play the underlying political differences already noted above, and the public involvement causing the domestic politics aspect to become highly prominent and thus further complicating efforts at a Community energy policy.

45

The causes of this politicization are not difficult to identify. They lie in a combination of long-term trends involving the change from an economy based on domestic coal to one based on imported oil, and in the crisis events of the October War with its attendant oil embargo, supply fears, and fourfold rise in oil prices. As Raymond Vernon has pointed out, the oil issue has been in the realm of high politics before (for example, for the United States at the time of the Mexican and Iranian oil nationalizations), but the size of the stakes has never been so great. In any case, a European energy policy which had previously seemed narrowly technical in scope suddenly became highly politicized, with consequences making it hard for the Europeans to achieve any progress.

Regional political unity in Europe has always been subject to the effects of politicization, though these interventions can also work to increase integration, particularly at times when integration presents a means to solve national problems.[132] Major advances (as opposed to lesser incremental developments) in European integration have thus ultimately depended upon prior political decisions much more than upon the operation of some quasi-automatic, neo-functional, invisible hand.[133] Significant progress, as in the creation of the European Coal and Steel Community, decisions by Prime Ministers Macmillan, Wilson, and Heath to press for British membership, and the Hague and Paris summits, has resulted from such direct political intervention, but so has severe obstruction or disruption, as in the defeat of the European Defense Community, the successive de Gaulle vetoes of British entry, de Gaulle's opposition to majority voting leading to the Luxembourg Agreement of 1966, and the Wilson Government's decision to renegotiate Britain's terms of EEC membership. In this light, the energy crisis clearly fits into the category of disruptive rather than supportive intervention.

Politicization transferred European energy policy from the realm of energy ministries and European integration offices to the desks of Foreign and Prime Ministers and somewhat decreased the direct policy influence of transnational actors, even though they continued to play a significant role. Decision criteria now became more explicitly based on national political and economic interests, and national differences thereby came to the fore, occasioning both bilateral competitive deals and broader multilateral responses beyond the EC level. In one sense this result is not surprising. Among decision-making elites interviewed by this author a few months prior to the crisis, politicians gave replies markedly less oriented toward European unity than did civil servants. On a composite European unity index, scaled from +5 to -5, civil servants scored +3.35 and politicans +2.14.[134] As the principal decisions about energy policy shifted from an administrative to a political realm, there was thus a decreased probability of a European response to the crisis.

Above all, however, politicization brought to the foreground the entire question of U.S.-European relations on both the economic and security planes, highlighting both the differences of outlook among the Europeans and the existing asymmetries in the relationship between America and Europe. Politicization also strengthened the linkages between economic and security policy which the U.S. had previously been less successful in asserting and which many of the Europeans had resisted during a period of growing Atlantic antagonism over monetary, economic, and trading matters.

b. Asymmetries and Linkages

Prior to the energy crisis there existed a widening belief that Europe had become an economic superpower comparable to the United States. To be sure, Europe was not a military superpower, but this seemed increasingly irrelevant at a time when international economic issues were growing in importance and impact. Given the existence of the Soviet-American nuclear balance, détente, and a lessening of direct superpower confrontation, it is not surprising that economic issues could be regarded as increasingly central to world politics and the source of growing division between the U.S. and Europe over trade, payments, investment, multinational corporations, and the like. Indeed, those European elites dealing with economic and monetary affairs displayed an increased sense of creating European unity not perhaps in opposition to the Americans but at least not in partnership with them.

In fact, some of these assumptions rested on a fragile basis of asymmetrical interdependence. While the Nine and the U.S. were interdependent in their economic and monetary structures, the Europeans remained much more vulnerable to the costs of disruption or change than did the Americans. A far higher proportion of European economic activity was based on foreign trade or was vulnerable to changes in monetary flows and exchange rates. Further, the Europeans remained dangerously vulnerable because of their high dependence on imports of energy and raw materials. Although the United States had experienced a recession, a long-term balance of payments deficit which precipitated successive dollar devaluations, and a growing balance of trade deficit, its economy remained far less vulnerable because of its greater size and lesser dependence on foreign economic activity. By contrast, the Europeans were not only vulnerable but also lacked a sufficiently strong centralized decision-making machinery to deal effec-

47

tively with these issues. Thus, the foreign economic policies of the EC were largely ad hoc in nature, the Commission was hobbled in its actions by having to wait for unanimous decisions by the Council of Ministers, and no sufficiently effective coordinating body existed within the Community. Moreover, the existing intergovernmental coordination of foreign policy through the Davignon Committee remained separate from the EC's own institutions.[135]

As Keohane and Nye have observed, power derives from asymmetrical interdependence, that is, a situation in which one side is more dependent or vulnerable to changes in a mutual relationship. As the efficacy of force decreases, threats to state autonomy may therefore shift from the area of military security to that of economic interdependence.[136] Indeed, the manipulation of this economic interdependence may become increasingly important and the use of force less so; alternatively, a state may be tempted to use its lesser military dependence to gain economic advantage.[137]

The impact of the energy crisis on Europe reflected asymmetries of power which had been previously latent. In their dealings with the United States, as well as with the oil producers, the Nine were vulnerable; in both sets of relationships the Europeans found themselves at a disadvantage, and in the Atlantic case, this enabled the Americans to exert linkage between economic issues and questions of American leadership and security commitments.[138] In a sense, energy itself became a kind of security issue, initially when the supply of oil was a matter of national survival and later when the ability to pay for this oil became a question of almost comparable gravity. In the words of a Dutch official, "The greatest threat to Europe's security is economic and it starts with oil."[139] Thus, with their economic and security interests threatened, economically vulnerable, and lacking a sufficient institutional basis to respond, the Europeans faced strong pressures to rely on United States leadership in energy and related economic matters as well as to pay greater heed to American political desires. As British Foreign Minister James Callaghan was soon to observe, "We repudiate the view that Europe will emerge only after a process of struggle against America." Instead he would urge the "fullest and most intimate cooperation" with the U.S.[140]

For the U.S., particularly the State Department, this presented a welcome opportunity to counter the increasing drift of the Community away from the United States, a movement which had been taking place under the impact of international economic issues, transnational forces, and the effects of détente. The U.S. had become impatient with bearing the security costs of alliance leadership in the face of declining linkage to economic and political questions. Indeed, the State Department and Secretary Kissinger had made it clear that further American support for European unity would only be forthcoming for a Europe which would work closely with the U.S. on economic and alliance issues, and that European unity could not be an end in itself.

The reassertion of American leadership was facilitated by an underlying European consensus on military security which had not been seriously eroded despite several years of European-American antagonism. At the height of these differences over non-military issues even the French had explicitly reaffirmed the need for a continued American troop presence in Europe in order to maintain an East-West military balance, although they tenaciously resisted efforts at being tied to American leadership on non-military issues. Thus, to the extent that the energy crisis came to be perceived in security terms, albeit broader and not primarily military ones, it tended to reinforce Atlantic ties by reemphasizing a realm in which these links were more substantial.

There are at least two strands of evidence for this conclusion. First, political elites and observers among the Nine widely believe that Secretary Kissinger and the U.S. successfully exerted pressure for linkage between a continued American military commitment to Europe and the acceptance of American leadership on energy policy. In the case of the West Germans, some Europeans have even characterized this pressure as "blackmail," evidentially because Kissinger, at the Washington Energy Conference, told German Foreign Minister Walter Scheel that the U.S. would reconsider the presence of U.S. troops in Germany unless the Europeans supported establishment of the Energy Coordinating Group and rejected the French position.[141] At the Washington meeting Kissinger met individually with each of the other European foreign ministers in order to convey a comparable message. Secretary of Defense James Schlesinger also expressed warnings about the possible removal of U.S. troops and President Nixon publicly threatened that unless the Europeans behaved the troops might be withdrawn.[142] Certainly the Nine recognized their dependence on the American commitment. Even at the December 1973 Copenhagen summit at which they had sought to appease the Arabs but remained divided among themselves, the final communiqué acknowledged that there was "no alternative" to the security provided by U.S. nuclear weapons and American forces in Europe.[143] Indeed, a year after the onset of the crisis, the Martinique accord between Presidents Ford and Giscard d'Estaing embodied a pledge to maintain continued close defense relations as members of the Atlantic Alliance. And, at the Washington Conference itself, the pull of this security tie helped to secure the agreement of all

49

but France to support the U.S.-sponsored Energy Coordinating Group, precisely the kind of outcome which the Nine, under French prodding, had explicitly ruled out before the meeting.

Yet another source of evidence may be found in European elite attitudes prior to the crisis. In essence, while those dealing with economic and monetary policy tended not to favor close Atlantic ties, persons who specialized in broader foreign policy matters and those dealing with military affairs displayed markedly greater support for close European-American relations. On a composite Atlantic Index scaled from +5 to -5, elites specializing in economic and monetary policy scored a mere +0.02, while the figure for foreign policy specialists was +1.20 and that for military policy elites was +1.87.[144] To the extent that crisis events displaced consideration of energy policy from a primarily economic and technical sphere and brought it to the attention of more political and security oriented elites, there was thus an increased probability that those dealing with the subject would be more Atlanticist in their response.

All these factors made the Europeans more receptive toward energy and other measures, including establishment of the IEA, which reinforced the European-American relationship and the reasserted predominance of the U.S. within it. In this context, the earlier view that military issues had become less crucial to international politics is not necessarily wrong. Nuclear parity, a lessening of bipolarity, and the increased salience of international economic issues are virtually self-evident and they imply a certain de-emphasis of traditional issues of force. But the lessons of the energy crisis do indicate that security must not be defined in narrowly military terms. In this sense the traditional distinctions between security and non-security issues have become less useful, and it is no longer appropriate always to equate security with military power and non-security issues with economic matters.[145] In the present case, European security was in a significant sense dependent upon oil and the awareness of this security dependence caused energy to be dealt with in a politicized manner; it also reinforced a multilateral response based on close ties to the United States.

c. Problems of Domestic Politics

Domestic politics have often been significant in shaping European international economic and foreign policies, but the politicization of the energy problem in the sense of involvement by the broader public made domestic considerations an even greater influence in affecting European responses to the energy crisis. To be sure, it is not only Western Europe which exhibits this pattern; the international energy policy of the Nixon and Ford administrations was also shaped by an innate political animus against rationing, planning, and interference with the size and gasoline consumption of automobiles — as well as by patterns of mass consumption based on low-cost energy — all of which sharply limited the possibilities for significant decreases in U.S. oil consumption.

In the European case, domestic political instabilities, problems of leadership, economic difficulties, inflation, unemployment, and the like all affected policy outcomes, but with different results in each of the principal European countries. In West Germany, increasingly the most important member of the Community, the desire to limit the amount of budgetary contributions to the EC, particularly in the absence of significant progress toward European unity, set limits on the Community's available resources. Further, Germany's domestic politics — as well as her geographical and international position — dictated the maintenance of strong Atlantic security links, thus tending to limit any drift of the Nine as a bloc toward a fundamental division with the U.S. Britain presents an even more noteworthy case of domestic political constraints. Conservative and Labour governments obstructed a common energy policy in order to bargain for a more generous European regional policy and later for renegotiation of Britain's membership. In both instances, the actions were taken because the European issues involved substantial domestic payoffs. Ironically, EC regional policy and hence the satisfaction of British political needs was limited by German reluctance to contribute substantially greater sums for this purpose — as a response to domestic political concerns in the Federal Republic. In the case of Italy, domestic political instability, severe economic problems (inflation, balance of payments deficits, unemployment), administrative chaos, and the presence of a powerful domestic Communist Party made that country a difficult partner for any grouping. Finally, in France, pressures from Gaullists, Communists, and parts of the Socialist Party made it politically costly for the governments of Pompidou and Giscard to contemplate close Atlantic ties or even a European energy policy which did not keep its distance from the Americans.

Consequently, the major EC members were unable to reach agreement on any significant Community policy which would not somewhere encounter grave domestic obstacles. Indeed, just as the Community found itself squeezed from above and below by responses to the energy crisis which were often bilateral or multilateral, so individual governments experienced conflicting pressures and demands. The governments were, on the one hand, impeded by international and transnational forces and, on the other, by increased domestic demands and problems as a result of a worsening international situation and pre-

51

existing constraints. These external impacts not only took the form of economic pressures (inflation caused by increased oil prices, soon to be accompanied by recession, as purchasing power was siphoned off by these energy costs) and of political problems (*vis-à-vis* the Middle East, Atlantic policy, etc.); they also threatened to have disruptive social implications (hoarding, as during the British coal strike, or the anti-Semitic implications of Saudi Arabian discrimination against European journalists of Jewish origin and Kuwaiti opposition to "Jewish banks" in London and Paris).[146]

Faced with an escalation of external and domestic pressures, accountable to national parliaments, parties, pressure groups, and electorates, urgently needing to find solutions to economic and political problems, and often lacking solid majorities of support, European governments would seek immediate short-term solutions. In an earlier period, domestic politics had made it convenient to support advances in European unity, for example at the 1969 Hague and 1972 Paris summits. Now, when domestic pressures, markedly intensified by international energy and economic problems and transnational forces, led the Nine in divergent directions, the Community's cohesiveness suffered.

4 AFTERWORD: FIVE PROPOSITIONS

The case of Europe in the energy crisis is of considerable interest in itself, but it also suggests five propositions of wider relevance and applicability.

Proposition 1: The politicization of international economic relations is increasing.

With the lessening of the rigid bipolarity which characterized the East-West conflict and Cold War, traditional international issues of force, borders, and armed conflict have — at least for a while — become a little less significant than before, while matters of international economics now occupy a prominent place on foreign affairs agendas. This shift has been stimulated by widespread economic development and the growth of a highly open and interdependent international economic system allowing for huge increases in trade, investment, monetary flows, technological innovation, and the operation of multinational corporations. But while international affairs have thus become more and more a matter of economics, these economic relations have, because of their intrinsic importance, become increasingly politicized themselves.

Although the heightened importance of economic issues in international (as well as domestic) politics has long been self-evident in Europe and Japan, elite and public awareness of this shift developed in the United States largely after Nixon's dramatic economic measures of August 1971 and the 1973-74 energy crisis. Secretary Kissinger, himself increasingly criticized for his lack of economic knowledge, has been moved to observe that "economic issues are turning into central political issues,"[147] and indeed Kissinger went out of his way to secure membership on the Ford Administration's Economic Policy Board, a body charged with responsibility for overseeing all U.S. foreign and domestic economic policy.

This politicization of international economic relations has taken place in terms both of increased attention by elite decision-makers and in greater public awareness of these issues. There have been several consequences of this. On the one hand, elite involvement has tended to lessen the traditional isolation of economic activity from foreign policy and domestic politics.[148] Decisions involving international economic activity now involve wider participation and the consideration of additional criteria. This has significant consequences for the role of transnational actors, particularly multinational corporations and international organizations, by lessening their impact on such decisions as well as their independence.

In addition, the existence of involvement by the broader public places greater constraints upon domestic decision-makers. Whether this is good or bad is another matter. While it may impede swift decision-making by complicating the criteria of choice and limiting the freedom of action of various participants, this type of politicization does introduce an element of political responsibility or accountability which scarcely exists elsewhere in the international economic sphere. To the extent that decisions are removed from determination primarily by transnational actors, or from government bureaucracies whose chief constituencies are these actors or other foreign and domestic pressure groups, and placed into a wider political realm, there exists at least the possibility — though by no means the certainty — that decisions may be taken with reference to broader public interest criteria. On the other hand, the introduction of these political and symbolic criteria may make it harder to achieve international agreements on economic problems which are the mutual concern of interdependent countries in both the developed and under-developed worlds.

While it is possible that this recent politicization is only a cyclical rather than a progressive phenomenon, and that the handling of international economic questions need not remain highly politicized in the future, there are reasons to believe that the change is more than temporary. Internationally, problems of economic development, resource and energy availability, technology, trade, monetary flows, foreign assistance, investment, and the activity of multinational corporations all imply a sustained political dimension. At the domestic level, increased governmental responsibility for economic performance (employment, price stability, prosperity, energy supplies, environment, and the like) similarly dictates a continued or expanded political treatment of international economic questions.

The consequences of this politicization lead, in turn, to a second proposition.

Proposition 2: Politicization tends to re-establish the primacy of traditional, unitary, state-to-state relations.

The literature on international economic relations and transnationalism has called attention to the manner in which traditional conceptions of international relations — which confined them to

FIGURE 1
Traditional, Transnational, and Transgovernmental Relations

Actors

	States as Actors	Non-state Actors
Centralized	Traditional inter-state relations A	B Transnational
Decentralized	C Transgovern-mental	D Transnational/ Transgovern-mental

States as Units

relations between states, each acting as a centralized unit — were seriously oversimplified. Instead, many significant international activities are undertaken by non-state, i.e., "transnational," actors or by government bodies which are not subject to direct control by the central government. In short, international affairs involves far more than the activities or influences of foreign ministries or the offices of presidents and prime ministers dealing directly with one another across national boundaries. Keohane and Nye's conception of this is illustrated in Figure 1, in which the traditional conception of international relations is designated by box A, but boxes B, C, and D symbolize important areas of transnational and transgovernmental activity which fall outside the traditional state-to-state sphere. The transnational literature has performed a valuable service in calling attention to the increased diversity of international actors as well as to the tendency of an enlarged agenda of international affairs to include issues previously regarded as strictly economic or technical. It has also helped to make people more aware of the blending of foreign and domestic politics and economics, the increased difficulties of states in maintaining foreign policy coherence, and the greater linkage among issue areas.

A significant question here is to delineate the conditions under which international relations take place under the transnational/transgovernmental conditions of Boxes B, C, and D rather than the traditional, Box A, framework. In fact, politicization appears to cause a reassertion of the primacy of traditional, unitary, state-to-state relations. The type of politicization inherent in increased attention by elite governmental decision-makers appears to lessen the scope for non-state, transnational actors. And politicization in the form of a broadened political arena seems to stimulate the involvement of centralized governmental machinery because of the political prominence of the questions involved and the need for presidents and prime ministers to reassert their control of sensitive areas for which they are held accountable by their electorates, public opinion, and the media. To be sure, there are issues which may become politicized in the sense of key decision-maker participation (e.g., oceans policy) without broader public involvement, and which may therefore leave open the possibility of transgovernmental relations between competing sub-national bureaucracies and interests. But in the energy case, politicization in both senses of the term dictated a reassertion of direct, centralized state-to-state dealings.

Such changes do not preclude the involvement of transnational actors, but they imply that these actors, for example the multinational corporations, may become more explicitly the instruments of state policy as governments seek to gain or regain authority in newly politicized spheres of international activity.

Proposition 3: Transnational forces do not necessarily pull advanced countries closer together, particularly those cooperating on a regional basis.

This proposition follows from the politicization of international economic relations (Proposition 1) and the reinvolvement of national governments in this activity (Proposition 2). The growth of interdependence, the rise of non-state and de-centralized governmental actors, and the enhanced importance of economic and resource questions may give rise to new and intensely divisive issues even as older traditional disputes among the developed countries, involving military forces, borders, colonies, and the like become less prominent or even disappear altogether.

The political intensity with which some of the newer issues may come to be regarded and the pressure on national decision-makers to find solutions to pressing economic and resource problems do not necessarily dictate cooperative regional patterns of response. As we have seen, pressures from the energy crisis overloaded the institutions and capabilities of the European Community and produced a fragmented pattern of policy response. By contrast, cooperation at a broader level involving the OECD, IEA, and — in particular — the United States was relatively more successful. A great deal thus depends on *which* groups of advanced countries one has in mind. It is conceivable that transnational pressures may work to pull together broader multilateral groupings but not narrower regional ones. Indeed, some would argue that conditions of interdependence can be coped with successfully only on the broadest basis of cooperation among the developed industrial states or even on a world level but that regional groupings are outmoded as vehicles for managing problems of energy, payments, trade, inflation, aid, and the like. Certainly an important conclusion of this study is that among the available multilateral approaches to the energy crisis, the European level proved to be too narrow or weak an area for effective action. By contrast, the OECD-IEA grouping offered a more successful means for coping. Consequently, one of the most interesting questions suggested by this case is not so much the old issue of national versus EC measures, but of an EC versus OECD-IEA format. It is certainly conceivable that responses at both the EC and wider levels will have continued or even enhanced importance.

In any case, evidence based on the European experience is ambiguous. On the one hand, had no multinational oil companies existed to allocate petroleum during the crisis, there might have been an even greater rupture within the Community and among the OECD countries. In this sense, prominent transnational actors did act to pull the advanced countries closer together. On the other hand, the resource and financial pressures of the crisis unleashed a scramble of competitive bidding for oil and for bilateral deals to manage balance of payments costs. Also in a negative light we have the observation of President Giscard d'Estaing that it is sometimes easier for France to deal with the Soviet Union than with the U.S. because relations with the latter often take place on a nongovernmental (i.e. transnational) basis:

> With the USSR, we deal only on a government-to-government basis. Many of our relations with the U.S. are non-governmental — on economics, finance, aviation, and so on. So its harder to keep an equality of relations.[149]

We also know that national behavior in a crisis situation tends to become less cooperative, witness the U.S. soybean export ban of mid-1973, the

57

10% import surcharge of August 1971, and various Italian, French, and British measures to deal with temporary monetary problems by means of exchange controls and trade barriers. It may be true that in non-crisis situations transnational forces and actors tend to pull advanced countries together, but the same factors also place increased pressure and responsibility for performance on national governments. This leads in turn to another proposition.

Proposition 4: In the face of increased national and transnational pressures governments will tend to use whatever works, on a pragmatic-instrumental rather than affective basis.

The growth of international economic relations and their politicization as well as that of domestic economic policy create intensified pressures on national governments to produce results. In consequence, various courses of action, whether bilateral, regional, or multilateral, tend increasingly to be seen as means for coping with national problems more than as desirable ends in themselves. Donald Puchala's study of the politics of rule enforcement in the EC finds that the Community is more and more regarded in such a light,[150] and the present analysis of Europe in the energy crisis indicates a similar conclusion. Puchala found that for France, in the case of the Common Agricultural Policy, and for Italy in labor migration, the EC could be used for domestic problem solving; by contrast, in the present case, the IEA, rather than the EC, offers a means of solving national problems on energy and is thus likely to prove a principal vehicle for most of the Nine in dealing with the problems of energy. In short, the pressures engendered by transnational and domestic economic realities predispose governments to pursue courses of action in the service of national interests and requirements rather than on the basis of institutional loyalty.

Proposition 5: Transnationalism and interdependence make it more difficult for governments to solve problems on a national basis, but some of the same factors also increase obstacles to multilateral solutions.

In an increasingly interdependent world, national governments face increased pressures from above and below. The international environment and transnational forces and actors may create or exacerbate domestic economic problems. Yet these same conditions make it increasingly difficult for governments to solve such problems on their own. The growth of the welfare state and of managed economies in the developed non-communist world has caused a huge expansion in the scope of state responsibility. Governments are thus required to cope with an expanded agenda of domestic problems and with the effects of international and transnational forces upon their local economies. Yet they find their problem-solving abilities constrained wherever they turn. Their ability to cope with domestic problems is limited by the extent to which these are caused or influenced externally (e.g., inflation, recession, unemployment, energy, even pollution). On the other hand, intensified domestic pressures upon governments to solve problems also make it harder for them to deal with these difficulties on a multilateral basis because of the pressure for short-term solutions and direct benefits.

It has already been noted how the United Kingdom twice obstructed EC energy policy because of a desire to bring about a substantial regional policy that would result in important domestic payoffs within Britain. Domestic pressures, however, prevented the German Federal Government from approving increased financing for such a policy.

More broadly, because governments in the developed world are increasingly perceived as responsible for economic and social well-being, they face increased pressures from a multiplicity of groups and parties. Governments are more and more obliged to intervene when problems arise, even though they are increasingly vulnerable to the effects of an international environment which they do not control and whose effects they often cannot cope with unilaterally or sometimes even multilaterally. The result is that problems become harder to solve, and decreased success in the domestic arena may undermine popular confidence in the government of the day as well as among its decision-makers.

At the same time, the increased importance of national pressures is not necessarily undesirable. In a normative sense, the national level is virtually the only place where matters of political responsibility, values, priorities, resource allocation, and the classic political questions of who gets what can be raised effectively. As the alternative to relegating these questions to the market sector or a kind of invisible hand at the national or multilateral level, the fact of government response to transnational, international, and domestic pressures embodies a certain value.

In the final analysis, governments may have a greater need to seek solutions to national problems at the multilateral level, even while constrained in these efforts by the effects of their domestic difficulties. The increased availability and even efficacy of organizations such as the IEA, as well as the manifest inadequacy of many types of national response, may well suggest a possible long-term trend in this direction.

Notes

1. Samuel Huntington, Joseph S. Nye, and Raymond Vernon have previously made this argument.

1a. The "Nine" refers to the member governments of the European Community (Britain, France, the Federal Republic of Germany, Italy, the Netherlands, Belgium, Denmark, Ireland, Luxembourg). The "Eight" is the same group exclusive of France.

2. Henri Simonet, quoted in *New York Times*, May 31, 1974.

3. A discussion of "politicization" follows in Part 3. The distinction between the "technical" and "political", or between "low" and "high" politics has received ample scholarly treatment elsewhere. *Viz.* the works of Stanley Hoffmann, Ernst Haas, Joseph Nye, and Robert Keohane. See also Robert J. Lieber, "Interest Groups and Political Integration: British Entry Into Europe," *American Political Science Review*, Vol. LXVI, No. 1 (March 1972), pp. 56-57.

4. For a useful survey of energy in this and the subsequent period see D. Swann, *The Economics of the Common Market* (Harmondsworth, Gt. Britain: Penguin, 1970), especially pp. 104ff., and John Nielsen, "Power: Society's Most Consumed Product," *European Community* No. 164 (April 1973), pp. 8-11.

5. Nielsen, p. 8.

5a. Romano Prodi and Alberto Clô, "Europe", in *Daedalus*, Fall 1975, p. 93.

6. "Coal" is used here to designate all solid fuel (including peat and lignite). 1960 and 1970 figures are for OECD Europe. Source: Organisation for Economic Cooperation and Development, *Oil, The Present Situation and Future Prospects* (Paris 1973), p. 265. 1950 figure from Nielsen, p. 8.

7. Vincent Roberts, "Chronology of a Crisis: Energy Shortage Hits Europe," *European Community*, No. 173 (February 1974) p. 10. The OEEC included most countries of Western Europe in its membership.

8. Between 1957 and 1966, EC coal production fell by 12%, the number of miners by 23%, and the number of pits by 42%. By 1965, EC coal was more expensive ($16.68 per ton) than imported U.S. coal ($14.20) or imported crude oil ($16.40). Source: ECSC, 1967, in Swann, *op. cit.*, pp. 106-108. During the 1960s, the number of coal miners employed among the Nine fell from 1.6 million to 615,000. Prodi and Clô, p. 92.

9. European Community. Commission, "First Guidelines for a Community Energy Policy" (December 18, 1968). Supplement to *Bulletin of the European Communities*, No. 12, 1968, p. 7.

10. John Walsh, "European Community Energy Policy: Regulation or Mainly Information?" *Science*, Vol. 184 (June 14, 1974), pp. 1158-1161.

11. Jack Hartshorn, "Europe's Energy Imports," in Max Kohnstamm and Wolfgang Hager, (eds.), *A Nation Writ Large? Foreign-Policy Problems Before the European Community* (N.Y.: Halsted Press, John Wiley & Sons, 1973), p. 111.

12. European Communities. Commission. "Necessary Progress in Community Energy Policy," (Communication from the Commission to the Council forwarded on October 13, 1972), in *Bulletin of the European Communities*, Supplement 11/72, p. 15.

13. *Ibid.*, p. 14.

14. European Communities. Commission. "Guidelines and Priority Actions Under the Community Energy Policy," (Communication from the Commission to the Council, April 27, 1973), in *Bulletin of the European Communities*, Supplement 6/73, p. 5.

15. "The main object here is to ensure effective competition and movement within the Community." *Ibid.*, p. 6.

16. Louis Turner, "Politics of the Energy Crisis," *International Affairs* (London), Vol. 50, No. 3, pp. 407-409.

17. *The Times* (London), January 23, 1973.

18. *The Times* (London), January 3, 1973.

19. *Le Monde* (Paris), July 25, 1973.

20. *Le Monde* (Paris), June 21, 1973.

21. *Time*, January 3, 1972.

22. See, *e.g.*, Robert J. Lieber, "Expanded Europe and the Atlantic Relationship," in Frans A. M. Alting von Geusau, (ed.), *The External Relations of the European Communities: Perspectives, Policies and Responses* (Farnborough, Gt. Britain: D. C. Heath Ltd., 1974), pp. 57-59; also Pierre Hassner, "L'Europe de la guerre froide à la paix chaude," *Defense Nationale* (March 1973), pp. 35-54; Stanley Hoffmann, "Weighing the Balance of Power," *Foreign Affairs*, Vol. 50, No. 4 (July 1972, pp. 618-43), and "Will the Balance Balance at Home," *Foreign Policy*, No. 7 (Summer 1972), pp. 60-86; Philip Windsor, ". . . But Europe Shouldn't," *Foreign Policy*, No. 8 (Fall 1972), pp. 92-99; and Alastair Buchan, "A World Restored?" *Foreign Affairs*, Vol. 50, No. 4 (July 1972), pp. 644-659.

23. Zbigniew Brzezinski, "The Balance of Power Delusion," *Foreign Policy*, No. 7 (Summer 1972), pp. 54-59.

24. *The Economist*, February 9, 1974, p. 180. Also Tad Szulc, "Is He Indispensable? Answers to the Kissinger Riddle," *New York Magazine*, July 1, 1974, p. 33, cited in Robert Pfaltzgraff, "The Middle East Crisis: Implications for the European-American Relationship," paper presented at the 1974 Annual Meeting of the American Political Science Association, Chicago, Illinois, August 29-September 2, 1974, p. 6.

25. Pfaltzgraff, *op. cit.*, p. 7.

26. *New York Times*, November 13, 1973.

27. *Le Monde*, November 2, 1973.

28. James O. Goldsborough, "France, the European Crisis and the Alliance," *Foreign Affairs*, Vol. 52, No. 3 (April 1974), p. 538.

29. Françoise de la Serre, "L'Europe des Neuf et le Conflit Israelo-Arabe," *Revue Française de Science Politique*, Vol. XXIV, No. 4 (August 1974), p. 805.

30. Walter Laqueur, *New York Times Magazine*, January 20, 1974, cited in Pfaltzgraff, *op. cit.*, p. 13.

31. "Le dossier arabe sur les Pays-Bas," *Maghreb-Machrek*, January-February 1974, pp. 13-17, quoted in de la Serre, p. 804. Cf. the statement of French Foreign Minister Michel Jobert, "Attempting to return home does not constitute unprovoked aggression." *Le Monde*, October 10, 1973, quoted *ibid.*

57. Turner, p. 412. He finds British-French expediency inglorious but effective in persuading the Arabs to restore oil cuts quietly.

58. Edward L. Morse has dealt with some of these elements in a lucid analysis. I particularly agree with his treatment of domestic factors. See "The New Europe: A Unified Bloc or Blocked Unity?" Princeton University, Woodrow Wilson School of Public and International Affairs, April 1974, mimeographed, pp. 30ff. On Germany, see Helmut Schmidt, "The Struggle for World Product," *Foreign Affairs*, Vol. 52, No. 3 (April 1974), p. 451.

59. European Communities. Commission. "Report on the Present or Foreseeable Impact of the Energy Supply Situation on Production, Employment, Prices, the Balance of Payments, and the Monetary Reserves," Brussels, Sec (74) 247, January 30, 1974.

60. European Communities. Commission. "Declaration on the State of the Community." January 31, 1974. See *Bull. EC* 1-1974.

61. "Collaboration with the U.S. in the Field of Energy," Communication from the Commission to the Council. Brussels, January 9, 1974. Sec (74) 68-E.

62. Quoted in *New York Times*, February 11, 1974.

63. *Bull. EC* 2-74, pp. 1201-1204.

64. *L'Express* (Paris), No. 1176. January 21-27, 1974.

65. Quoted in *New York Times*, February 12, 1974.

66. Quoted in *New York Times*, February 13, 1974.

67. *New York Times*, February 15 and 18, 1974.

68. Quoted in *New York Times*, February 15, 1974.

69. November 13, 1973.

70. *European Community*, No. 176, May 1974, p. 5.

71. *The Times* (London), March 6, 1974.

72. Interview in *Le Monde*, March 8, 1974, quoted in *The Times* (London), March 9, 1974.

73. *Ibid.*

74. Address to Association Belgo-Americaine, Brussels, March 21, 1974. Reprinted in *Bull. EC* 3-74, pp. 5-8.

75. "Relations Between the Community and the Energy-producing Countries," (Communication from the Commission to the Council). Brussels, January 23, 1974. Com. (74) 90, p. 8.

76. "Guidelines and Priority Actions Under the Community Energy Policy." (Communication from the Commission to the Council, April 27, 1973). *Bull. EC.* Supp. 6/73, p. 6.

77. One French Foreign Ministry official privately termed the PLO question merely an "excuse" for delaying the talks. (Interview with the author, Paris, March 21, 1975.)

In June 1975 the Europeans and Arabs managed to finesse the Palestinian issue by agreeing that the Community and the Arab League would each be represented by a single delegation whose composition would be of its own choosing. There was also a brief Arab expression of pique over the signing of an Israeli-EEC association agreement. The talks finally opened on June 10, 1975.

78. A meeting of EC Commissioners Henri Simonet and Claude Cheysson with the OAPEC Secretary-general, Ali Attiga, did take place in Brussels on October 7-8, 1974, at which the participants agreed to develop regular contacts and exchange technical information on their activities. *European Community*, No. 182 (December 1974), p. 16.

32. *The Economist*, November 24, 1973.

33. *New York Times*, November 5, 1973.

34. *New York Times*, November 7, 1974.

35. Declaration of Principles of International Law Concerning Friendly Relations and Cooperation Among States in Accordance with the Charter of the United Nations: "No State may use or encourage the use of economic, political or any other type of measures to coerce another State in order to obtain from it the subordination of the exercise of its sovereign rights and to secure from it advantages of any kind." Richard N. Gardner, "The Hard Road to World Order," *Foreign Affairs*, Vol. 52, No. 3 (April 1974), p. 567.

36. Based on a sample of 13,000 people in the nine EC countries. See Martin U. Mauthner, "The Politics of Energy," *European Community* (March 1974), p. 13. According to a SOFRES poll, 70% of the French favored oil sharing. *L'Express*, November 12, 1973.

37. Louis Turner, "Politics of the Energy Crisis," p. 410.

38. See, *e.g.*, *The Economist*, November 24, 1973, and *New York Times*, November 20, 1973.

39. Jean de Lipkowski, State Secretary in the Foreign Ministry, quoted in *New York Times*, November 28, 1973.

40. *New York Times*, November 30, 1973.

41. *New York Times*, December 5, 1973.

42. *Ibid.*

43. *New York Times*, March 19 and March 23, 1974.

44. *New York Times*, July 11, 1974.

45. Geoffrey Chandler, "The Changing Shape of the Oil Industry," *Petroleum Review*, June 1974, quoted in Statement of Professor Robert B. Stobaugh, Harvard University Graduate School of Business Administration, before the Subcommittee on Multinational Corporations of the Committee on Foreign Relations, U.S. Senate, on Multinational Petroleum Companies and Foreign Policy. July 25, 1974, mimeographed, p. 14.

46. *The Economist*, "Multinational Business Report," December 1973, p. 1.

47. Stobaugh, *op. cit.*, pp. 10-11.

48. *Ibid.*, p. 11.

49. Turner, *loc. cit.*, pp. 410 ff.

50. *New York Times*, December 11, 1973.

51. Stobaugh, *op. cit.*

52. *Ibid.*, p. 3. On a country by country basis, the availability petroleum between December 1973 and March 1974 was lower than in the sa period of the previous year by 16% in the Netherlands, 12% in Germany in France, 1% in Britain, and was higher by 4% in Italy. Prodi and Clô, p.

53. *E.g.*, The German Federal Republic had higher oil stocks on ha December 15th (62 days) than on December 1st (59 days); the French por Havre received 25% more oil in November 1973 than in November 1972; I and Venezuela were believed to have increased production for the Neth *New York Times*, December 22, 1973.

54. *The Times* (London), March 7, 1974.

55. *The Economist*, January 26, 1974.

56. *Bulletin of the European Committees*, hereinafter cited as F 1974, pp. 1206-1212.

79. *The Economist* (London), September 21, 1974. For details of the proposal see "Towards a New Energy Policy Strategy for the European Community," memorandum submitted by the Commission to the Council, in *Bull. EC* 5-1974, pp. 1201 ff.

80. *The Economist* (London), July 27, 1974.

81. *The Times* (London), October 17, 1974. See also Henri Simonet, "Energy and the Future of Europe," *Foreign Affairs*, Vol. 53, No. 1 (April 1975), p. 456.

82. *New York Times*, November 29, 1974.

83. *Le Nouvel Observateur* (Paris), No. 538, March 3, 1975. At a Gaullist Party conference, former Premier Pierre Messmer warned against any agreement with the U.S. (implicitly, the IEA) which would return France to membership of a bloc. *New York Times*, December 16, 1974.

84. European Communities. Commission. "The Activities of the International Energy Agency" and "The Preparation of the Consumer/Producer Dialogue," (Commission Communication to the Council, Brussels, January 10, 1975), Com. (75) 5 final. Also see *Le Monde* (Paris), January 22, 1975.

85. European Communities. Commission. "Energy Questions to be Determined at the Community Level," (Communication of the Commission to the Council, Brussels, January 15, 1975), Com. (75) 6 final, Annex 1, p. 2.

86. *New York Times*, April 17, 1975, and *The Economist* (London), April 19, 1975.

87. *The Economist* (London), January 18, 1975.

88. While the evidence for this is quite dubious, the interpretation was given to the author by a knowledgeable and well-placed official at the French Foreign Ministry. (Interview, Paris, March 12, 1975). Another account indicating the French favored action to break the oil cartel and even suggested possible military action was reported by newspaper columnist Jack Anderson in late May 1974 but was denied by Secretary Kissinger and former UN Ambassador John Scali.

89. *The Guardian* (London), January 12, 1974.

90. See, *e.g.*, *New York Times*, January 29, 1974.

91. Richard Kershaw, *New Statesman* (London), March 22, 1974, p. 391.

92. "Relations Between the Community and the Energy-producing Countries," (Communication from the Commission to the Council). Brussels, January 23, 1974. Com. (74) 90, p. 3.

93. European Communities. Commission, "Towards a New Energy Policy Strategy for the Community," (Communication presented to the Council by the Commission on June 5, 1974). *Bull. EC.* Supplement 4/74.

94. *New York Times*, January 18, 1974.

95. *New York Times*, January 26, 1974.

96. Address to the Foreign Affairs Committee of the French National Assembly, October 10, 1974, quoted in *European Community*, November 1974.

97. Quoted in *New York Times*, February 17, 1974.

98. *The Times* (London), March 7 and September 26, 1974.

99. *New York Times*, June 28, 1974.

100. See, *e.g.*, *New York Times*, December 24 and 27, 1974.

101. Quoted in *New York Times Magazine*, December 15, 1974.

102. *The Economist* (London), September 7, 1974.

103. *The Economist* (London), January 11, 1975, and *New York Times*, January 4, 1975.

104. *The Economist*, January 25, 1975.

105. *New York Times*, November 13, 1973.

106. *The Times* (London), February 23, 1974.

107. *New York Times*, July 9, 1974.

108. *International Herald Tribune* (Paris), March 5, 1975.

109. Chef de Cabinet to one of the EC Commissioners, interviewed by the author, Brussels, March 20, 1975.

110. *Le Monde* (Paris), January 19-20, 1975.

111. On the logic of Prisoner's Dilemma see, e.g., Robert J. Lieber, *Theory and World Politics* (Cambridge, Mass.: Winthrop Publishers, 1972), pp. 28-32. A comparable logic is reflected in Jean-Jacques Rousseau's analogy of the stag hunt.

112. Henri Simonet, *Foreign Affairs*, April 1975, p. 458.

113. Walter J. Levy, "An Atlantic-Japanese Energy Policy," paper delivered at the Europe-America Conference, Amsterdam, March 1973. Also see *Foreign Policy*, No. 11 (Summer 1973), pp. 159-190.

114. Foreign Minister Jean Sauvagnargues, *Le Monde* (Paris), January 19-20, 1975.

115. "Assistant Secretary Enders Outlines Draft Agreement Reached by Energy Coordinating Group," *Department of State Bulletin*, Vol. LXXI, No. 1843, October 21, 1974, p. 526.

116. The initial formula of the ECG provided for each member state to have 3 votes plus a certain number of additional votes based on its oil consumption (for the U.S., 51 out of 100 total consumption votes). A weighted majority required 60% of total votes. Slightly different calculations for a weighted majority apply, depending on whether the embargo is general or selective against specific members. E.g., to delay a selective embargo requires the vote of 10 states regardless of weight. Draft of International Energy Program, *New York Times*, September 30, 1974.

117. Interview with the author, Paris, March 17, 1975.

118. *New York Times*, September 30, 1974.

119. Interviewed by Arnaud de Borchgrave, *Newsweek*, January 13, 1975.

120. European Communities. Commission. "Towards a New Energy Policy Strategy for the Community," (Communication presented to the Council by the Commission on June 5, 1974), *Bull. EC.* Supplement 4/74, p. 24. Also *The Times* (London) October 26, 1974, and *European Community*, No. 182 (December 1974), p. 18.

121. Henri Simonet, *Foreign Affairs*, (April 1975), p. 460.

122. OECD. *Energy Prospects to 1985*. Vol. I. Paris, 1974, p. 5.

123. Speech in Chicago, cited in *New York Times*, November 15, 1974.

124. Dankwart A. Rustow, "The Political Economy of Energy and the Euro-American System," outline prepared for Conference on Organizing the Euro-American System, (mimeographed). Arnoldsbain, January 27-30, 1975.

125. The wisdom of treating integration as a multidimensional phenomenon is widely accepted. The notion of examining separately institutional integration, social bases, and policy cooperation follows in part from the work of Donald Puchala, who has distinguished between the linking of peoples (community formation) and the linking of governments (political amalgamation), and of Joseph S. Nye, who has "dissected" regional integration into economic, social, and political integration, as well as differentiated between

institutional, policy, and attitudinal integration. See Puchala, "Integration and Disintegration in Franco-German Relations, 1954-1965," in *International Organization*, Vol. 34 (Spring 1970), pp. 184-85; Nye, *Peace in Parts: Integration and Conflict in Regional Organization* (Boston: Little Brown, 1971), pp. 27-49, and "The Political Context," in Lawrence B. Krause and Walter S. Salant (eds.), *European Monetary Unification and its Meaning for the United States* (Washington, D.C.: Brookings, 1973), pp. 40-41.

126. Lindberg's full scale of decision locus (based on an idea of William Riker) reads as follows:

1. Decisions are taken entirely in the European Community System.
2. Decisions are taken almost entirely in the European Community System.
3. Decisions are taken predominantly in the European Community System, but the nation-states play a significant role in decision-making.
4. Decisions are taken about equally in the European Community system and the nation-states.
5. Decisions are taken predominantly by the nation-states, but the European Community system plays a significant role in decision-making.
6. Decisions are taken almost entirely by the nation-state.
7. Decisions are taken entirely by the nation-states individually.

"The European Community as a Political System: Notes Toward the Construction of a Model," *Journal of Common Market Studies*, Vol. 5 (June 1967), p. 344. Also see the application to the EEC, Central America, and East Africa by J. S. Nye, "Comparative Regional Integration: Concept and Measurement," *International Organization*, Vol. 21 (1968), p. 870.

127. In essence, the assigning of values and the equal weighting of categories is based on the author's arbitrary judgment.

128. Henri Simonet, "Energy and the Future of Europe," *Foreign Affairs*, Vol. 53, No. 1 (April 1975), p. 452.

129. Cited in Martin Mauthner, "The Politics of Energy," *European Community*, No. 174 (March 1974), p. 13.

130. Lieber, "Interest Groups and Political Integration: British Entry Into Europe," *American Political Science Review*, Vol. LXVI, No. 1 (March 1972), pp. 56-57. Also see Stanley Hoffmann, "European Process at Atlantic Cross-purposes," *Journal of Common Market Studies*, (February 1965), p. 92.

131. This distinction is noted briefly in Keohane and Nye, "Transgovernmental Relations and International Organizations," paper presented at the 1974 Annual Meeting of The American Political Science Association, Chicago, August 29-September 2, 1974, pp. 25-27.

132. See Donald J. Puchala, "The Domestic Politics of Supranational Harmonization in the European Community," paper presented at the 1974 Annual Meeting of the American Political Science Association, Chicago, August 29-September 2, 1974.

133. Leon Lindberg and Stuart Scheingold draw a useful distinction between incremental progress in integration ("forward linkage growth") and major integrative advance requiring political intervention ("systems transformation"). *Europe's Would-Be Polity* (Englewood Cliffs: Prentice-Hall, 1970), p. 343.

134. For a more comprehensive explanation and discussion of this scale and the questions upon which it is based, see Robert J. Lieber, "European Elite Attitudes Revisited: The Future of the European Community and European-American Relations," in *British Journal of Political Science* (October 1975), pp. 335-352, and see especially table 9, p. 347.

135. See Robert McGheehan and Steven Warnecke, "Europe's Foreign Policies," *Orbis* (Winter 1974), pp. 1262-1265.

136. Robert O. Keohane and Joseph S. Nye, "World Politics and the International Economic System," in C. Fred Bergsten (ed.), *The Future of the International Economic Order: An Agenda for Research* (Lexington: D.C. Heath, Lexington Books, 1973), pp. 118-122.

137. Keohane and Nye, "International Interdependence and Integration," in Fred Greenstein and Nelson Polsby (eds.), *The Handbook of Political Science* (Reading, MA.: Addison-Wesley, 1975), and in Bergsten, *loc. cit.*, p. 120.

138. Keohane and Nye see a secular trend favoring increased linkages, in Bergsten, *loc. cit.*, p. 132.

139. Interview with the author, January 10, 1975.

140. Quoted in *The Guardian* (London), March 23, 1974.

141. This version of events is given by a well-placed French Foreign Ministry official who was sympathetic to the IEA. Interview with the author, Paris, March 12, 1975.

142. Speech in Chicago, March 15, 1974, cited in *The Economist* (London), March 23, 1974.

143. See *International Herald Tribune* (Paris), December 15/16, 1973.

144. See Lieber, "European Elite Attitudes Revisited," *loc. cit.*, p. 347.

145. I agree here with Frans A. M. Alting von Geusau in his criticisms of Seyom Brown's position. "Security problems in the Euro-American System," International Working Conference on Organizing the Euro-American System, January 27-30, 1975, mimeographed, p. 18.

146. Characteristically, the French government did not support French journalists denied visas to accompany government officials to Saudi Arabia. By contrast, the Dutch Prime Minister delayed a Middle East visit for several weeks until visas were granted to Jewish journalists accompanying his official party.

147. Quoted in *New York Times*, June 8, 1975. Similarly, the Trilateral Commission has observed that international economic relations will take on an increasingly political character. See "Energy: The Imperative for a Trilateral Approach." A Report of the Trilateral Task Force on the Political and International Implications of the Energy Crisis to the Executive Committee of the Trilateral Commission. Rapporteurs: John C. Campbell, Guy de Carmoy, Shinichi Kondo. Brussels, June 23-25, 1974. Keohane and Nye also argue that "We are now witnessing increasing politicization of international economic affairs." In Bergsten, *loc. cit.*, p. 118.

148. Keohane and Nye also see increased possibility for linkage between these issues. In Bergsten, *loc. cit.*, p. 131.

149. Quoted in *New York Times*, December 21, 1974.

150. Donald Puchala, *op. cit.* Also see "Domestic Politics and Regional Harmonization in the European Communities," *World Politics* Vol. XXVII, No. 4 (July, 1975), pp. 496-520.

BOOKS WRITTEN UNDER CENTER AUSPICES

The Soviet Bloc, Zbigniew K. Brzezinski (sponsored jointly with the Russian Research Center), 1960. Harvard University Press. Revised edition, 1967.

The Necessity for Choice, by Henry A. Kissinger, 1961. Harper & Bros.

Rift and Revolt in Hungary, by Ferenc A. Váli, 1961. Harvard University Press.

Strategy and Arms Control, by Thomas C. Schelling and Morton H. Halperin, 1961. Twentieth Century Fund.

United States Manufacturing Investment in Brazil, by Lincoln Gordon and Engelbert L. Grommers, 1962. Harvard Business School.

The Economy of Cyprus, by A. J. Meyer, with Simos Vassiliou (sponsored jointly with the Center for Middle Eastern Studies), 1962. Harvard University Press.

Entrepreneurs of Lebanon, by Yusif A. Sayigh (sponsored jointly with the Center for Middle Eastern Studies), 1962. Harvard University Press.

Communist China 1955-1959: Policy Documents with Analysis, with a foreword by Robert R. Bowie and John K. Fairbank (sponsored jointly with the East Asian Research Center), 1962. Harvard University Press.

Somali Nationalism, by Saadia Touval, 1963. Harvard University Press.

The Dilemma of Mexico's Development, by Raymond Vernon, 1963. Harvard University Press.

Limited War in the Nuclear Age, by Morton H. Halperin, 1963. John Wiley & Sons.

In Search of France, by Stanley Hoffmann *et al.*, 1963. Harvard University Press.

The Arms Debate, by Robert A. Levine, 1963. Harvard University Press.

Africans on the Land, by Montague Yudelman, 1964. Harvard University Press.

Counterinsurgency Warfare, by David Galula, 1964. Frederick A. Praeger, Inc.

People and Policy in the Middle East, by Max Weston Thornburg, 1964. W. W. Norton & Co.

Shaping the Future, by Robert R. Bowie, 1964. Columbia University Press.

Foreign Aid and Foreign Policy, by Edward S. Mason (sponsored jointly with the Council on Foreign Relations), 1964. Harper & Row.

How Nations Negotiate, by Fred Charles Iklé, 1964. Harper & Row.

Public Policy and Private Enterprise in Mexico, edited by Raymond Vernon, 1964. Harvard University Press.

China and the Bomb, by Morton H. Halperin (sponsored jointly with the East Asian Research Center), 1965. Frederick A. Praeger, Inc.

Democracy in Germany, by Fritz Erler (Jodidi Lectures), 1965. Harvard University Press.

The Troubled Partnership, by Henry A. Kissinger (sponsored jointly with the Council on Foreign Relations), 1965. McGraw-Hill Book Co.

The Rise of Nationalism in Central Africa, by Robert I. Rotberg, 1965. Harvard University Press.

Pan-Africanism and East African Integration, by Joseph S. Nye, Jr., 1965. Harvard University Press.

Communist China and Arms Control, by Morton H. Halperin and Dwight H. Perkins (sponsored jointly with the East Asian Research Center), 1965. Frederick A. Praeger, Inc.

Problems of National Strategy, ed. Henry Kissinger, 1965. Frederick A. Praeger, Inc.

Deterrence before Hiroshima: The Airpower Background of Modern Strategy, by George H. Quester, 1966. John Wiley & Sons.

Containing the Arms Race, by Jeremy J. Stone, 1966. M.I.T. Press.

Germany and the Atlantic Alliance: The Interaction of Strategy and Politics, by James L. Richardson, 1966. Harvard University Press.

Arms and Influence, by Thomas C. Schelling, 1966. Yale University Press.

Political Change in a West African State, by Martin Kilson, 1966. Harvard University Press.

Planning without Facts: Lessons in Resource Allocation from Nigeria's Development, by Wolfgang F. Stolper, 1966. Harvard University Press.

Export Instability and Economic Development, by Alasdair I. MacBean, 1966. Harvard University Press.

Foreign Policy and Democratic Politics, by Kenneth N. Waltz (sponsored jointly with the Institute of War and Peace Studies, Columbia University), 1967. Little, Brown & Co.

Contemporary Military Strategy, by Morton H. Halperin, 1967. Little, Brown & Co.

Sino-Soviet Relations and Arms Control, ed. Morton H. Halperin (sponsored jointly with the East Asian Research Center), 1967. M.I.T. Press.

Africa and United States Policy, by Rupert Emerson, 1967. Prentice-Hall.

Elites in Latin America, edited by Seymour M. Lipset and Aldo Solari, 1967. Oxford University Press.

Europe's Postwar Growth, by Charles P. Kindleberger, 1967. Harvard University Press.

The Rise and Decline of the Cold War, by Paul Seabury, 1967. Basic Books.

Student Politics, ed. S. M. Lipset, 1967. Basic Books.

Pakistan's Development: Social Goals and Private Incentives, by Gustav F. Papanek, 1967. Harvard University Press.

Strike a Blow and Die: A Narrative of Race Relations in Colonial Africa, by George Simeon Mwase, ed. Robert I. Rotberg, 1967. Harvard University Press.

Party Systems and Voter Alignments, edited by Seymour M. Lipset and Stein Rokkan, 1967. Free Press.

Agrarian Socialism, by Seymour M. Lipset, revised edition, 1968. Doubleday Anchor.

Aid, Influence, and Foreign Policy, by Joan M. Nelson, 1968. The Macmillan Company.

Development Policy: Theory and Practice, edited by Gustav F. Papanek, 1968. Harvard University Press.

International Regionalism, by Joseph S. Nye, 1968. Little, Brown, & Co.

Revolution and Counterrevolution, by Seymour M. Lipset, 1968. Basic Books.

Political Order in Changing Societies, by Samuel P. Huntington, 1968. Yale University Press.

The TFX Decision: McNamara and the Military, by Robert J. Art, 1968. Little, Brown & Co.

Korea: The Politics of the Vortex, by Gregory Henderson, 1968. Harvard University Press.

Political Development in Latin America, by Martin Needler, 1968. Random House.

The Precarious Republic, by Michael Hudson, 1968. Random House.

The Brazilian Capital Goods Industry, 1929-1964 (sponsored jointly with the Center for Studies in Education and Development), by Nathaniel H. Leff, 1968. Harvard University Press.

Economic Policy-Making and Development in Brazil, 1947-1964, by Nathaniel H. Leff, 1968. John Wiley & Sons.

Turmoil and Transition: Higher Education and Student Politics in India, edited by Philip G. Altbach, 1968. Lalvani Publishing House (Bombay).

German Foreign Policy in Transition, by Karl Kaiser, 1968. Oxford University Press.

Protest and Power in Black Africa, edited by Robert I. Rotberg, 1969. Oxford University Press.

Peace in Europe, by Karl E. Birnbaum, 1969. Oxford University Press.

The Process of Modernization: An Annotated Bibliography on the Sociocultural Aspects of Development, by John Brode, 1969. Harvard University Press.

Students in Revolt, edited by Seymour M. Lipset and Philip G. Altbach, 1969. Houghton Mifflin.

Agricultural Development in India's Districts: The Intensive Agricultural Districts Programme, by Dorris D. Brown, 1970. Harvard University Press.

Authoritarian Politics in Modern Society: The Dynamics of Established One-Party Systems, edited by Samuel P. Huntington and Clement H. Moore, 1970. Basic Books.

Nuclear Diplomacy, by George H. Quester, 1970. Dunellen.

The Logic of Images in International Relations, by Robert Jervis, 1970. Princeton University Press.

Europe's Would-Be Polity, by Leon Lindberg and Stuart A. Scheingold, 1970. Prentice-Hall.

Taxation and Development: Lessons from Colombian Experience, by Richard M. Bird, 1970. Harvard University Press.

Lord and Peasant in Peru: A Paradigm of Political and Social Change, by F. LaMond Tullis, 1970. Harvard University Press.

The Kennedy Round in American Trade Policy: The Twilight of the GATT? by John W. Evans, 1971. Harvard University Press.

Korean Development: The Interplay of Politics and Economics, by David C. Cole and Princeton N. Lyman, 1971. Harvard University Press.

Development Policy II — The Pakistan Experience, edited by Walter P. Falcon and Gustav F. Papanek, 1971. Harvard University Press.

Higher Education in a Transitional Society, by Philip G. Altbach, 1971. Sindhu Publications (Bombay).

Studies in Development Planning, edited by Hollis B. Chenery, 1971. Harvard University Press.

Passion and Politics, by Seymour M. Lipset with Gerald Schaflander, 1971. Little, Brown, & Co.

Political Mobilization of the Venezuelan Peasant, by John D. Powell, 1971. Harvard University Press.

Higher Education in India, edited by Amrik Singh and Philip Altbach, 1971. Oxford University Press (Delhi).

The Myth of the Guerrilla, by J. Bowyer Bell, 1971. Blond (London) and Knopf (New York).

International Norms and War between States: Three Studies in International Politics, by Kjell Goldmann, 1971. Published jointly by Läromedelsförlagen (Sweden) and the Swedish Institute of International Affairs.

Peace in Parts: Integration and Conflict in Regional Organization, by Joseph S. Nye, Jr., 1971. Little, Brown & Co.

Sovereignty at Bay: The Multinational Spread of U.S. Enterprise, by Raymond Vernon, 1971. Basic Books.

Defense Strategy for the Seventies (revision of *Contemporary Military Stategy*), by Morton H. Halperin, 1971. Little, Brown & Co.

Peasants Against Politics: Rural Organization in Brittany, 1911-1967, by Suzanne Berger, 1972. Harvard University Press.

Transnational Relations and World Politics, edited by Robert O. Keohane and Joseph S. Nye, Jr., 1972. Harvard University Press.

Latin American University Students: A Six Nation Study, by Arthur Liebman, Kenneth N. Walker, and Myron Glazer, 1972. Harvard University Press.

The Politics of Land Reform in Chile, 1950-1970: Public Policy, Political Institutions and Social Change, by Robert R. Kaufman, 1972. Harvard University Press.

The Boundary Politics of Independent Africa, by Saadia Touval, 1972. Harvard University Press.

The Politics of Nonviolent Action, by Gene E. Sharp, 1973. Porter Sargent.

System 37 Viggen: Arms, Technology, and the Domestication of Glory, by Ingemar Dörfer, 1973. Universitetsforlaget (Oslo).

University Students and African Politics, by William John Hanna. 1974. Africana Publishing Company.

Organizing the Transnational: The Experience with Transnational Enterprise in Advanced Technology, by M. S. Hochmuth, 1974. Sijthoff (Leiden).

Becoming Modern, by Alex Inkeles and David H. Smith, 1974. Harvard University Press.

The United States and West Germany 1945-1973: A Study in Alliance Politics, by Roger Morgan (sponsored jointly with the Royal Institute of International Affairs), 1974. Oxford University Press.

Multinational Corporations and the Politics of Dependence: Copper in Chile, 1945-1973, by Theodore Moran, 1974. Princeton University Press.

The Andean Group: A Case Study in Economic Integration among Developing Countries, by David Morawetz, 1974. M.I.T. Press.

Kenya: The Politics of Participation and Control, by Henry Bienen, 1974. Princeton University Press.

Land Reform and Politics: A Comparative Analysis, by Hung-chao Tai, 1974. University of California Press.

Big Business and the State: Changing Relations in Western Europe, edited by Raymond Vernon, 1974. Harvard University Press.

Economic Policymaking in a Conflict Society: The Argentine Case, by Richard D. Mallon and Juan V. Sourrouille, 1975. Harvard University Press.

New States in the Modern World, edited by Martin Kilson, 1975. Harvard University Press.

Revolutionary Civil War: The Elements of Victory and Defeat, by David Wilkinson, 1975. Page-Ficklin Publications.

Politics and the Migrant Poor in Mexico City, by Wayne A. Cornelius, 1975. Stanford University Press.

East Africa and the Orient: Cultural Synthesis in Pre-Colonial Times, by H. Neville Chittick and Robert I. Rotberg, 1975. Africana Publishing Company.

No Easy Choice: Political Participation in Developing Countries, by Samuel P. Huntington and Joan M. Nelson, 1976. Harvard University Press.

The Politics of International Monetary Reform - The Exchange Crisis, by Michael J. Brenner, 1976. Ballinger Publishing Co.

The International Politics of Natural Resources, by Zuhayr Mikdashi, 1976. Cornell University Press.

The Oil Crisis, edited by Raymond Vernon, 1976. W. W. Norton & Co.

Social Change and Political Participation in Turkey, by Ergun Ozbudun, 1976. Princeton University Press.

Harvard Studies in International Affairs*

[formerly Occasional Papers in International Affairs]

† 1. *A Plan for Planning: The Need for a Better Method of Assisting Underdeveloped Countries on Their Economic Policies*, by Gustav F. Papanek, 1961.

† 2. *The Flow of Resources from Rich to Poor*, by Alan D. Neale, 1961.

† 3. *Limited War: An Essay on the Development of the Theory and an Annotated Bibliography*, by Morton H. Halperin, 1962.

† 4. *Reflections on the Failure of the First West Indian Federation*, by Hugh W. Springer, 1962.

5. *On the Interaction of Opposing Forces under Possible Arms Agreements*, by Glenn A. Kent, 1963. 36 pp. $1.25.

† 6. *Europe's Northern Cap and the Soviet Union*, by Nils Orvik, 1963.

7. *Civil Administration in the Punjab: An Analysis of a State Government in India*, by E. N. Mangat Rai, 1963. 82 pp. $1.75.

8. *On the Appropriate Size of a Development Program*, by Edward S. Mason, 1964. 24 pp. $1.00.

9. *Self-Determination Revisited in the Era of Decolonization*, by Rupert Emerson, 1964. 64 pp. $1.75.

10. *The Planning and Execution of Economic Development in Southeast Asia*, by Clair Wilcox, 1965. 37 pp. $1.25.

11. *Pan-Africanism in Action*, by Albert Tevoedjre, 1965. 88 pp. $2.50.

12. *Is China Turning In?* by Morton Halperin, 1965. 34 pp. $1.25.

†13. *Economic Development in India and Pakistan*, by Edward S. Mason, 1966.

14. *The Role of the Military in Recent Turkish Politics*, by Ergun Özbudun, 1966. 54 pp. $1.75.

†15. *Economic Development and Individual Change: A Social-Psychological Study of the Comilla Experiment in Pakistan*, by Howard Schuman, 1967.

16. *A Select Bibliography on Students, Politics, and Higher Education*, by Philip G. Altbach, UMHE Revised Edition, 1970. 65 pp. $2.75.

17. *Europe's Political Puzzle: A Study of the Fouchet Negotiations and the 1963 Veto*, by Alessandro Silj, 1967. 178 pp. $3.50.

18. *The Cap and the Straits: Problems of Nordic Security*, by Jan Klenberg, 1968. 19 pp. $1.25.

19. *Cyprus: The Law and Politics of Civil Strife*, by Linda B. Miller, 1968. 97 pp. $3.00.

†20. *East and West Pakistan: A Problem in the Political Economy of Regional Planning*, by Md. Anisur Rahman, 1968.

†21. *Internal War and International Systems: Perspectives on Method*, by George A. Kelley and Linda B. Miller, 1969.

†22. *Migrants, Urban Poverty, and Instability in Developing Nations*, by Joan M. Nelson, 1969. 81 pp.

23. *Growth and Development in Pakistan, 1955-1969*, by Joseph J. Stern and Walter P. Falcon, 1970. 94 pp. $3.00.

24. *Higher Education in Developing Countries: A Select Bibliography*, by Philip G. Altbach, 1970. 118 pp. $4.00.
25. *Anatomy of Political Institutionalization: The Case of Israel and Some Comparative Analyses*, by Amos Perlmutter, 1970. 60 pp. $2.50.
26. *The German Democratic Republic from the Sixties to the Seventies*, by Peter Christian Ludz, 1970. 100 pp.
27. *The Law in Political Integration: The Evolution and Integrative Implications of Regional Legal Processes in the European Community*, by Stuart A. Scheingold, 1971. 63 pp. $2.50.
28. *Psychological Dimensions of U.S.-Japanese Relations*, by Hiroshi Kitamura, 1971. 46 pp. $2.00.
29. *Conflict Regulation in Divided Societies*, by Eric A. Nordlinger, 1972. 137 pp. $4.25.
30. *Israel's Political-Military Doctrine*, by Michael I. Handel, 1973. 101 pp. $3.25.
31. *Italy, NATO, and the European Community: The Interplay of Foreign Policy and Domestic Politics*, by Primo Vannicelli, 1974. 67 + x pp. $3.25.
32. *The Choice of Technology in Developing Countries: Some Cautionary Tales*, by C. Peter Timmer, John W. Thomas, Louis T. Wells, Jr., and David Morawetz, 1975. 114 pp. $3.45.
33. *The International Role of the Communist Parties of Italy and France*, by Donald L. M. Blackmer and Annie Kriegel, 1975. 67 + x pp. $2.75.
34. *The Hazards of Peace: A European View of Detente*, by Juan Cassiers, 1976. $6.95, cloth; $2.95, paper.
35. *Oil and the Middle East War: Europe in the Energy Crisis*, by Robert J. Lieber. 1976. $7.45, cloth; $3.45, paper.

*Available from Harvard University Center for International Affairs, 6 Divinity Avenue, Cambridge, Massachusetts 02138
†Out of print. Reprints may be ordered from AMS Press, Inc., 56 East 13th Street, New York, N.Y. 10003